Fractional Order Thinking in Exploring the Frontiers of STEM

Fractional Order Thinking (FOT) is about solving today's complex problems in the physical, social and life sciences using the tools of fractional order calculus (FOC). Very soon after Mandelbrot introduced the fractal paradigm into the scientific lexicon it was shown that the integer order calculus (IOC) could not describe the dynamics of fractal processes. A new kind of calculus was required to construct the equations of motion for fractal dynamic processes which turned out to be fractional order Hamiltonian equations (FOHEs). The FOHEs are one tool in the FOC toolbox which concerns how to apply the operators of differentiation and integration of non-integer orders. Rejecting the fractional calculus is equivalent to saying there are no numbers between neighboring integers.

In this book series, we explore the core motivation of fractional calculus by first showing the "core motivation" of IOC invented by Newton and Leibnitz, the fundamental ideas of which can be traced back to the time of Heraclitus of Ephesus. Our ultimate message is that the "IOC" is driven by "the desire and the need" for the "quantification of changes" based on energy gradients in complex dynamic networks, whereas "FOC" is driven by "the desire and the need of understanding complexity" based on information gradients in those same networks. In science, technology, engineering, and mathematics (STEM), metaphorically speaking, there is plenty of room between the integers to enable the archiving of better than the best modelling, control performance, robustness, resilience, and even intelligence.

There is an increasing interest in fractional order dynamic systems (FODS) and controls in the recent research literature, not only because of their novelty but also due to their practical applications. But to accomplish all of this requires a new way of thinking, the Fractional Order Thinking (FOT) referred to above, which in turn must be preceded by a new STEM curriculum. This series will offer a unique platform to demonstrate such additional benefits in our improved understanding of complexity and stochastic dynamics via FOT.

This exciting book series features:

- A forum demonstrating the good consequences of using fractional order thinking (FOT).
- Fractional calculus concepts are presented in the context of complex systems characterization while assuming minimal background in math and physics.
- A broad audience including professionals across many fields, the general public and courses in colleges and even high schools.
- Wide STEM topics ranging from batteries to medicine with a writing style easy to follow.
- Short and inexpensive books of 120-150 pages main text that can be written and read in a reasonable amount of time.

Please contact the series editors, Bruce J. West (North Carolina State University, USA) and YangQuan Chen (University of California Merced), and Taylor & Francis Publisher Lian Sun (Lian.Sun@informa.com), if you have an idea for a book for the series.

Titles in the series currently include:

Fractional Calculus for Skeptics I
The Fractal Paradigm
Bruce J. West and YangQuan Chen

On the Fractal Language of Medicine
Bruce J. West and W. Alan C. Mutch

Fractional Order Intelligent Modeling for Lithium-Ion Batteries
Theory and Practice
YaNan Wang and YangQuan Chen

Fractional Random Vibrations I
Theories
Ming Li

Fractional Random Vibrations II
Applications
Ming Li

Fractional Order Intelligent Modeling for Lithium-Ion Batteries

This book focuses on fractional order (non-integer order) modeling (FOM) techniques coupled with deep neural network-based intelligent modeling methods for lithium-ion batteries (LIBs) and battery management systems (BMS) in general. It provides the first one-stop resource on FOM for LIBs with case studies using real operational data sets.

With the rapid growth of electric vehicles and energy storage systems, battery technology has become critical to global energy solutions. *Fractional Order Intelligent Modeling for Lithium-Ion Batteries: Theory and Practice* aims to provide several accurate and effective intelligent modeling algorithms for the next generation of advanced BMS. Key topics include intelligent battery modeling, fractional-order modeling, physics-informed machine learning, state estimation, and degradation analysis. By integrating AI and physics-informed machine learning techniques with fractional-order modeling methods, this book presents several innovative solutions for next-generation battery management systems.

This title will serve as an invaluable resource for researchers and advanced students in the fields of transportation, energy storage, and power systems, as well as those studying electric vehicles, control theory, machine learning, and fractional calculus-based modeling.

YaNan Wang is currently an assistant professor and a member of the Low-Carbon Powertrain Systems Research Lab at Beijing University of Technology, China. Her research focuses on AI-driven battery intelligent management and safety evaluation for power batteries, addressing critical issues such as fast degradation and fault diagnosis.

YangQuan Chen is a professor at the University of California Merced, US. His research interests include mechatronics for sustainability, digital twins, small multi-UAV, and applied fractional calculus. His recent publications with CRC Press include: *Fractional Calculus for Skeptics I: The Fractal Paradigm* and *Fractional Calculus for Skeptics II: Quantifying Roughness*.

Fractional Order Intelligent Modeling for Lithium-Ion Batteries

Theory and Practice

YaNan Wang and YangQuan Chen

CRC Press
Taylor & Francis Group
Boca Raton London New York

CRC Press is an imprint of the
Taylor & Francis Group, an **informa** business

First edition published 2026
by CRC Press
2385 NW Executive Center Drive, Suite 320, Boca Raton FL 33431

and by CRC Press
4 Park Square, Milton Park, Abingdon, Oxon, OX14 4RN

CRC Press is an imprint of Taylor & Francis Group, LLC

ISBN: 978-1-041-13269-1 (hbk)
ISBN: 978-1-041-13663-7 (pbk)
ISBN: 978-1-003-67090-2 (ebk)

DOI: 10.1201/9781003670902

Typeset in Latin Modern font
by KnowledgeWorks Global Ltd.

Publisher's note: This book has been prepared from camera-ready copy provided by the authors.

To our mentors and families.

Contents

Preface

Fractional calculus (FC) is about differentiation and integration of non-integer order. Fractional-order thinking is thinking in-between integers when using integer order (traditional) calculus for analysis, modeling, and control of real-world complex systems. Denying fractional calculus is like saying there are no non-integers in between integers. Fractional calculus has been proposed and applied to various mechanical, chemical, and electrical fields. It has been demonstrated in the literature that by using fractional-order models, we can better characterize the complex system dynamics or behaviors, and by using fractional-order controls, we can better control the complex dynamic systems than the best performance the integer order counterparts could offer. This "better than the best" implication of applying fractional calculus is particularly attractive to engineers and scientists in general.

This text aims at providing an integrated overview of fractional-order modeling for the lithium-ion battery (LIB). The fractional-order modeling idea was first introduced to electrochemical elements or electroanalysis in 1973 by Professor Keith Oldham.[1,2] In fact, as an electrochemist, Prof. Oldham is perhaps the first to connect fractional calculus to the Warburg diffusion element, an equivalent electrical circuit component that models the diffusion process in dielectric spectroscopy. This element, now called the "fractional-order element", is named after the German physicist Emil Gabriel Warburg. Today, in electrochemistry, Warburg impedance and the more general constant phase element (CPE), widely used as building blocks in modeling of electrochemical energy storage systems such as LIB, are a focus in this book. Note that the CPE idea was first introduced in the 1940s by Kenneth Stewart Cole who utilized the CPE to model the electrical impedance of biological materials, particularly in studies related to cell membranes. It is interesting to note that Kenneth Stewart Cole and Robert Hugh Cole proposed the famous Cole–Cole relaxation model, often used to describe dielectric relaxation in polymers, which is a fractional-order model and a departure from the standard Deybe relaxation model (integer order), which can much better capture the algebraic decay (inverse power decay) than the exponential decay. It is interesting to note that Prof. Oldham interacted with the second author in 2004 at the first IFAC Workshop on Fractional Derivatives and Applications (FDA'2004) in Bordeaux, France, and the paper title was *Robust PID Controller Autotuning with a Phase Shaper*, where the key idea was also linked to CPE but in a controller design context explicitly linked to fractional calculus in the

---

[1]Keith Oldham and Jerome Spanier. *The fractional calculus theory and applications of differentiation and integration to arbitrary order*, volume 111, Elsevier, 1974.

[2]Keith B. Oldham. *Semiintegral electroanalysis. analog implementation.* Analytical Chemistry, 45(1):39–47, 1973.

paper. Prof. Oldham liked the talk and gave the presenter his business card with a handwritten note on the back.

Over the past 50 years, fractional-order capacitors, fractional-order circuits, and fractional-order machine learning have been gradually used for LIB modeling and battery management systems (BMS). In the past two decades, the papers using the fractional calculus idea for LIB modeling and BMS have been increasing in numbers in quite diverse outlets. For a beginner, it is hard to capture the essence of the large number of scattered papers. This text tries to take an overview on the main aspects of fractional calculus for LIB and provide several fractional-order modeling algorithms developed by the authors for LIBs in a single volume for the beginners to start with.

Part of this text is extracted from the first author's PhD dissertation and post-doc report. Part of the book contents is also based on the first author's one-year stay with the Mechatronics Embedded Systems and Automation (MESA) Lab of the University of California Merced (2018–2019) focusing on LIB modeling, machine learning, and microwave radar sensing (Walabot). The roadmap of this book is shown in Figure 1.

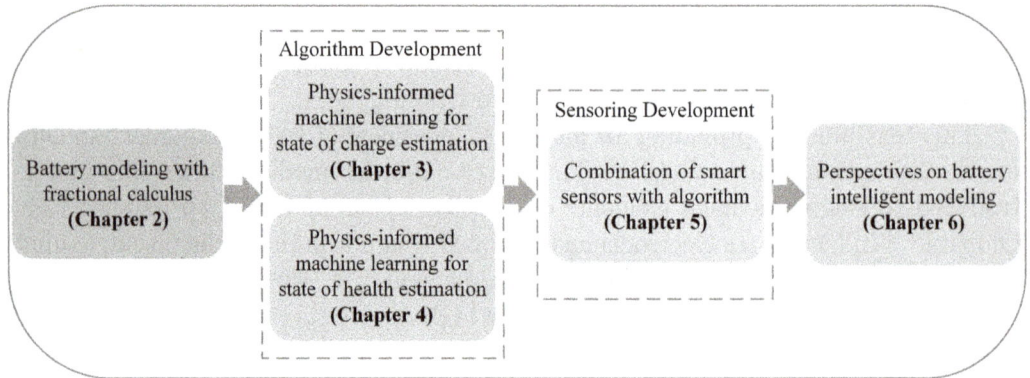

Figure 1   Book roadmap.

The structure and chapter arrangement of this book are as follows.

Chapter 1 is an introduction to the book. In Chapter 2, an overview of state-of-the-art fractional-order modeling and estimation for LIBs is presented. In Chapter 3, we introduce physics-informed conditions for battery state estimation. In Chapter 4, we introduce fractional-order modeling for battery capacity estimation. In Chapter 5, we introduce intelligent modeling for battery with smart sensors. In Chapter 6, we introduce perspectives on intelligent fractional-order modeling during whole battery life-cycle. Finally, Chapter 7 concludes the book with discussions of future research directions with some take-home messages.

YaNan Wang would like to thank Prof. Minggao Ouyang and his research team members from Tsinghua University for their support and help. She wishes to thank Prof. Changwei Ji and his research team members from Beijing University of Technology for the support. Dr. Wang also wishes to thank the funding support by National Natural Science Foundation of China under Grant No. 62103220. Finally, Dr. Wang wishes to thank the China Scholarship Council (CSC) for funding her one-year exchange PhD studentship at the MESA Lab of the University of California, Merced.

YangQuan Chen wishes to thank the funding support of the Center for Methane Emission Research and Innovation (CMERI) through the Climate Action Seed Funds grant (2023–2026) at the University of California, Merced.

Last but not least, the authors would like to thank Ms. Xiaoyin Feng, Editorial Assistant, Routledge and CRC Press/Taylor & Francis Group, for her professional coordination of the book project; and Ms. Lian Sun, Publisher, Taylor & Francis, for her professional vision and effort in establishing this Fractional Order Thinking in Exploring the Frontiers of STEM book series to which this book belongs.

Beijing, China, August 2025   *YaNan Wang*
Merced, CA, USA, August 2025   *YangQuan Chen*

Authors

YaNan Wang (PhD/lecturer/master's supervisor) is a member of the Low-Carbon Powertrain Systems Research Team at Beijing University of Technology. She earned her PhD from Beijing Institute of Technology in 2020, then worked as a post-doctoral fellow in Tsinghua University in 2020–2023. She has published over 20 first/corresponding-author papers in *Nature Communications* and other international journals/conferences and has obtained five authorized patents. She leads the National Natural Science Foundation of China (NSFC) Youth Project, has participated in five national/provincial-level projects as a core member in the past five years, and directed four enterprise-commissioned projects. Her team focuses on AI-driven battery intelligent management and safety evaluation for power batteries, addressing critical issues like fast degradation and fault diagnosis. She has long-term collaboration with Prof. Minggao Ouyang's team at Tsinghua University. She also cooperates with domestic automakers (BAIC, SGMW, Hyundai) and battery manufacturers (CATL, Keyi Power) on electric vehicle powertrain management.

YangQuan Chen earned his PhD from Nanyang Technological University, Singapore in 1998. He had been a faculty member of the Electrical Engineering Department at Utah State University (USU) from 2000–2012. He joined the School of Engineering, University of California, Merced (UCM) in 2012 teaching mechatronics, engineering service learning, and unmanned aerial systems for undergraduates; fractional order mechanics, nonlinear controls, and advanced controls: optimality and robustness for graduates. His research interests include mechatronics for sustainability, cognitive process control, small multi-UAV-based cooperative multi-spectral "personal remote sensing", applied fractional calculus in controls, modeling, and complex signal processing; distributed measurement and control of distributed parameter systems with mobile actuator and sensor networks. He was listed in Highly Cited Researchers by Clarivate Analytics from 2018–2021. He received Research of the Year awards from USU (2012) and UCM (2020). In 2018, for his cumulative work in drones, Dr. Chen won the Senate Distinguished Scholarly Public Service Award, which recognizes a faculty member who has energetically and creatively applied his or her professional expertise and scholarship to benefit the local, regional, national, or international community. He received the Control Engineering Practice Best Journal Paper Award at the IFAC World Congress 2011 in Milan, Italy. He is proud of receiving a Counselor Award from the Utah State University Student Branch of IEEE for "explaining human relationship using control theory".

Abbreviations

There are numerous abbreviations used throughout the book. Books from various fields may have abbreviations that use the same letters but have different meanings. So to avoid confusion, we list in alphabetical order the abbreviations used.

3D	Three-Dimensional
AGI	Artificial General Intelligence
AI	Artificial Intelligence
BMS	Battery Management System
CC-CV	Constant-Current Constant-Voltage
CNN	Convolutional Neural Network
CPE	Constant Phase Element
DL	Deep Learning
DOC	Depth of Charge
DT	Digital Twin
ECM	Equivalent Circuit Model
EIS	Electrochemical Impedance Spectrum
EKF	Extended Kalman Filter
EL	Extreme Learning
EOL	End of Life
ESS	Energy Storage System
EV	Electric Vehicle
FC	Fractional Calculus
FL	Federated Learning
FOGD	Fractional-Order Gradient Descent
FO-KF	Fractional-Order Kalman Filter
FOM	Fractional-Order Modeling
FORNN	Fractional-Order Recurrent Neural Network
fPIRNN	Physics-Informed Recurrent Neural Network with Fractional-Order Constraints
FUDS	Federal Urban Driving Conditions
GA	Genetic Algorithm
GD	Gradient Descent
GDm	Gradient Descent with Momentum
GRU	Gated Recurrent Unit

HEV	Hybrid Electric Vehicle
HPPC	Hybrid Pulse Power Characteristic
IC	Incremental Capacity
ICA	Incremental Capacity Analysis
IoT	Internet of Things
KF	Kalman Filter
LAM	Loss of Active Material
LDA	Linear Discriminant Analysis
LFP	$LiFePO_4$
LIB	Lithium-Ion Battery
LLI	Loss of Lithium-Ion
LSTM	Long Short-Term Memory
ML	Machine Learning
MSE	Mean Square Error
MSS	Mean Sum of Square
NAG	Nesterov Accelerated Gradient
NCM	Ni-CO-Mn
NLP	Natural Language Processing
NN	Neural Network
OCV	Open Circuit Voltage
P2D	Pseudo-Two-Dimensions
PCA	Principal Component Analysis
PDE	Partial Differential Equation
PDF	Probability Distribution Function
PIML	Physics-Informed Machine Learning
PINN	Physics-Informed Neural Network
PIRNN	Physics-Informed Recurrent Neural Network
PNGV	Partnership for a New Generation of Vehicles
PSO	Particle Swarm Optimization
RF	Radio Frequency
RNN	Recurrent Neural Network
RUL	Remaining Usage Life or Remaining Useful Life
SEI	Solid Electrolyte Interphase
SGD	Stochastic Gradient Descent
SMO	Sliding Mode Observer
SOC	State of Charge
SOH	State of Health
SPM	Single Particle Model
SSM	State Space Model
SVD	Singular Value Decomposition
SVM	Support Vector Machine
UKF	Unscented Kalman Filter
VR	Virtual Reality

CHAPTER 1

Introduction

1.1 FRACTIONAL CALCULUS

Fractional calculus, a mathematical theory capable of characterizing anomalous phenomena or behaviors of complex systems, has been driving extensive research efforts in complex system modeling and prediction over the past decades. Fractional calculus extends classical integer-order calculus to non-integer orders, resulting in a fractional-order operator, that is, fractional derivative and fractional integral [6]. Fractional calculus enables more precise modeling of complex systems with memory effects and multiscale dynamics [7].

The mathematical concept of fractional-order calculus, initially proposed over 300 years ago, is now developed with three widely adopted modern formulations or definitions: Riemann-Liouville (R-L), Grünwald-Letnikov (G-L), and Caputo definitions [12, 4]. However, due to implementation challenges in the discretization of fractional-order calculus [9], critical advancements have only emerged in recent decades, then gradually been applied in engineering research domains [8]. Fractional-order state space models (SSMs), fractional-order PID controllers, constant phase elements (CPEs), and fractional-order fluid dynamics, to name a few, are widely used in current research [10, 5]. Recent computational advances facilitate the applications of fractional calculus in control theory, material science, and electrochemical systems such as lithium-ion batteries (LIBs) [19]. Fractional calculus for LIBs can effectively characterize complex behaviors and heterogeneous processes through fractional differential equations [13, 2, 22].

1.2 LITHIUM-ION BATTERY

In response to the growing energy demands across various applications, batteries have emerged as highly efficient, high-density energy storage devices. They serve as primary components in solar, wind, and hydroelectric energy storage systems (ESS), and also as power sources in transportation field, maintaining their status as a persistent research focus across relevant scientific and technological fields [21, 3]. LIBs excel among various battery types, including lead-acid, zinc-based, nickel-based, and hydrogen-oxygen fuel cells, because LIBs have the advantages of high operating

DOI: 10.1201/9781003670902-1

voltage, high energy density, relatively low self-discharge rate, high-current load tolerance, and minimal maintenance requirements.

As shown in Figure 1.1, a battery contains an anode, a cathode, a separator, an electrolyte, and two current collectors (positive and negative). Both the anode and the cathode contain lithium ions. Through the separator, the electrolyte transports positively charged lithium ions from the anode to the cathode and vice versa. The mobility of lithium ions generates free electrons in the anode, which in turn produces a charge in the positive current collector. The current then flows from the positive current collector through the load device to the negative current collector. The separator prevents electrons from flowing through the battery. While the battery is draining and supplying an electric current, the anode discharges lithium ions to the cathode, resulting in an electron flow from one side to the other. When charging the battery, the converse occurs: the cathode releases lithium ions and the anode receives them. The two current collectors in anode and cathode are always made of aluminum (Al) and copper (Cu), respectively. The positive electrode (anode) operates at a high potential, where the copper foil is easily oxidized; thus, it can only be the cathode collector. Since aluminum exhibits a higher oxidation potential and the dense oxide film on the aluminum foil surface provides effective protection to the underlying aluminum, aluminum is chosen as the anode collector.

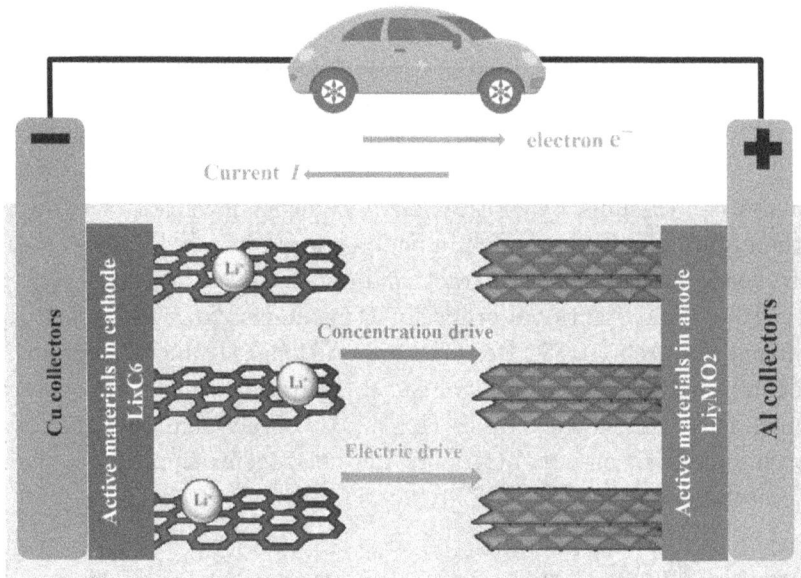

Figure 1.1 Working fundamentals of lithium-ion batteries (LIBs).

Current research directions for LIBs can be categorized into three main directions. The first direction focuses on range extension, lifespan enhancement, efficiency improvement, energy optimization, recycling of secondary LIBs, and charging technology upgrades. The second direction addresses the development of fast-charging technology, higher energy density, and heating solutions under low-temperature conditions for LIBs. The third direction targets safety management and cost control for future high-density and high-energy LIBs. These directions rely fundamentally on acquiring and analyzing extensive LIB data to extract key parameters for monitoring and control. In the era of big data, deep learning (DL) and machine learning (ML) algorithms offer intelligent and data-driven optimization strategies to advance LIB research across all three main above-mentioned directions [18].

1.3 FRACTIONAL-ORDER MODELING

Based on fractional calculus, fractional-order modeling is proposed by constructing ordinary or partial differential equations (PDEs) with fractional orders [14]. Fractional-order modeling employs non-integer-order differential operators to mathematically represent systems exhibiting memory-dependent dynamics, multiscale phenomena, and non-exponential or algebraic (inverse power) relaxation processes. This approach extends classical integer-order calculus frameworks, enabling precise characterization of anomalous diffusion, viscoelastic material responses, and electrochemical interfacial behaviors. In energy storage systems such as LIBs, fractional-order models can capture diffusion and dynamics through elements like constant phase elements (CPE) in impedance spectroscopy and fractional-order state space representations for capacity degradation analysis [15, 16, 17], as shown in Figure 1.2.

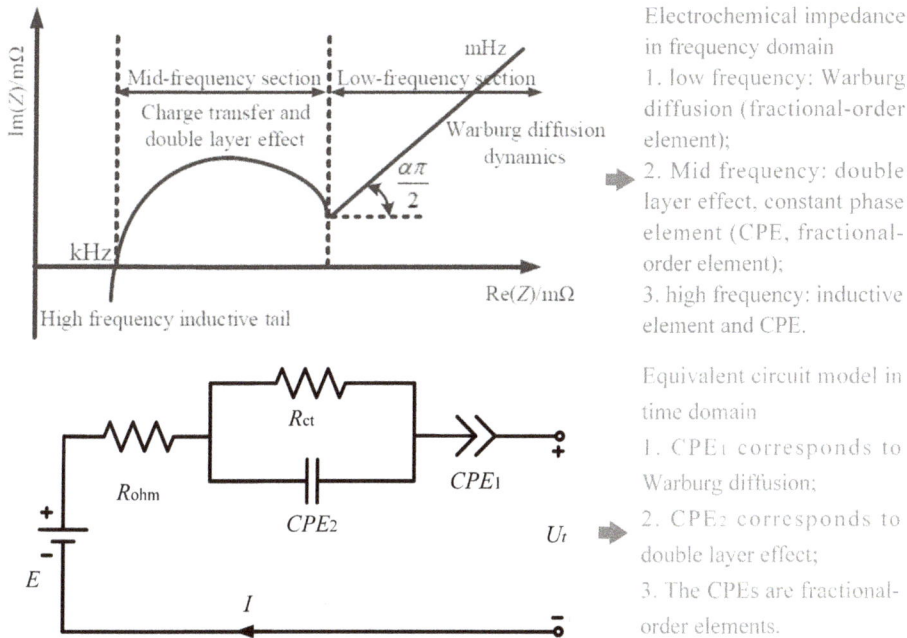

Figure 1.2 An example of fractional-order modeling for battery both in frequency domain and time domain.

Key implementations include fractional-order PID controllers for enhanced system regulation, fractional viscoelastic constitutive equations for polymer analysis, and fractional diffusion equations for heterogeneous reaction kinetics. The methodology demonstrates superior performance in describing frequency-dependent phase shifts, power-law relaxation patterns, and distributed time constants inherent to complex physical systems [1, 20]. Recent computational advancements in numerical discretization, parameter identification, and stability analysis have expanded fractional calculus applications to bioengineering, thermal management, and renewable energy systems, particularly in modeling aging mechanisms of electrochemical devices and multiphysics-coupled transport processes [11, 22].

1.4   MAIN CONTRIBUTIONS

The major contributions of this monograph are briefly summarized in the following list.

1. Introduction of fractional-order modeling for battery, especially intelligent modeling by machine learning.

2. The combination of battery mechanism with machine learning, and the applications of smart sensors for battery.

3. Physics-informed neural network for battery estimation designed and verified, including fractional-order gradients and fractional-order constraints.

4. Both experimental data and practical running data of electrical vehicles (EVs) applied to test the proposed algorithms.

5. Perspectives on battery intelligent management and modeling.

1.5   BOOK ORGANIZATION

This monograph is structured as follows. The research motivations and contributions are introduced in Chapter 1.

In Chapter 2, we give an overview of state-of-the-art fractional-order modeling and estimation for LIBs, including electrochemical models, equivalent circuit models (ECMs), and estimation methods for state of charge (SOC), state of health (SOH), and remaining useful life (RUL).

In Chapter 3, we introduce physics-informed conditions for battery state estimation. Combined with fractional-order characteristics of LIBs, fractional-order gradient and fractional-order constraint are proposed. Then, a physics-informed recurrent neural network (PIRNN) is constructed for battery modeling. SOC estimation is conducted by the proposed PIRNN algorithm. Experiment results, analysis and comparison are presented in detail.

In Chapter 4, we introduce fractional-order modeling for battery capacity estimation. Based on fractional-order information of battery degradation, both physics-informed inputs and structure of neural network are designed. Experimental results with realistic EVs running data are presented and analyzed.

In Chapter 5, we introduce intelligent modeling for battery with smart sensors. Smart perception is reviewed and two kinds of sensors are presented as examples. One is a kind of micro-wave sensor called Walabot for battery voltage classification, the other is a kind of millimeter wave sensor called ImageVK74 for noncontact battery capacity estimation.

In Chapter 6, we introduce perspectives on intelligent fractional-order modeling during the whole battery life-cycle. Intelligent management system with fractional calculus for battery modeling is provided. Opportunities and challenges for future work are also discussed.

Finally, Chapter 7 concludes the monograph with discussions of future research directions with some take-home messages for every chapter.

1.6 RESULTS REPRODUCIBILITY

Codes for the methods presented in this monograph will be published here: `https://github.com/Wangking7ai/CRC-Book-Series-FOT4STEM-volume-3.git` so that interested readers can reproduce the results presented in this book more easily.

Bibliography

[1] Tohid Soleymani Aghdam, Seyed Mohammad Mahdi Alavi, and Mehrdad Saif. Structural identifiability of impedance spectroscopy fractional-order equivalent circuit models with two constant phase elements. *Automatica*, 144:110463, 2022.

[2] Masoud Alilou, Hatef Azami, Arman Oshnoei, Behnam Mohammadi-Ivatloo, and Remus Teodorescu. Fractional-order control techniques for renewable energy and energy-storage-integrated power systems: A review. *Fractal and Fractional*, 7(5):391, 2023.

[3] Xinyuan Bao, Liping Chen, António M Lopes, Xin Li, Siqiang Xie, Penghua Li, and YangQuan Chen. Hybrid deep neural network with dimension attention for state-of-health estimation of lithium-ion batteries. *Energy*, 278:127734, 2023.

[4] Paul L Butzer and Ursula Westphal. An introduction to fractional calculus. In *Applications of Fractional Calculus in Physics*, pages 1–85. World Scientific, 2000.

[5] X Dingyu and Bai Lu. *Fractional Calculus*. Springer, 2024.

[6] Rudolf Gorenflo and Francesco Mainardi. *Fractional Calculus: Integral and Differential Equations of Fractional Order*. Springer, 1997.

[7] Rudolf Hilfer. *Applications of Fractional Calculus in Physics*. World Scientific, 2000.

[8] Changpin Li and Fanhai Zeng. *Numerical Methods for Fractional Calculus*. Chapman and Hall/CRC, 2015.

[9] Ch Lubich. Discretized fractional calculus. *SIAM Journal on Mathematical Analysis*, 17(3):704–719, 1986.

[10] Concepción A Monje, YangQuan Chen, Blas M Vinagre, Dingyu Xue, and Vicente Feliu-Batlle. *Fractional-Order Systems and Controls: Fundamentals and Applications.* Springer Science & Business Media, 2010.

[11] K Dhananjay Rao, Allamsetty Hema Chander, and Subhojit Ghosh. Robust observer design for mitigating the impact of unknown disturbances on state of charge estimation of lithium iron phosphate batteries using fractional calculus. *IEEE Transactions on Vehicular Technology*, 70(4):3218–3231, 2021.

[12] Bertram Ross. The development of fractional calculus 1695–1900. *Historia Mathematica*, 4(1):75–89, 1977.

[13] JinPeng Tian, Rui Xiong, WeiXiang Shen, and FengChun Sun. Fractional order battery modelling methodologies for electric vehicle applications: Recent advances and perspectives. *Science China Technological Sciences*, 63:2211–2230, 2020.

[14] Yanan Wang, Yangquan Chen, and Xiaozhong Liao. State-of-art survey of fractional order modeling and estimation methods for lithium-ion batteries. *Fractional Calculus and Applied Analysis*, 22(6):1449–1479, 2019.

[15] Yanan Wang, Xuebing Han, Feng Dai, Jie Li, Daijiang Zou, Languang Lu, Yangquan Chen, and Minggao Ouyang. Fractional order backpropagation neural network for battery capacity estimation with realistic vehicle data. In *2022 18th IEEE/ASME International Conference on Mechatronic and Embedded Systems and Applications (MESA)*, pages 1–6. IEEE, 2022.

[16] Yanan Wang, Xuebing Han, Dongxu Guo, Languang Lu, Yangquan Chen, and Minggao Ouyang. Physics-informed recurrent neural network with fractional-order gradients for state-of-charge estimation of lithium-ion battery. *IEEE Journal of Radio Frequency Identification*, 6:968–971, 2022.

[17] Yanan Wang, Xuebing Han, Dongxu Guo, Languang Lu, Yangquan Chen, and Minggao Ouyang. Physics-informed recurrent neural networks with fractional-order constraints for the state estimation of lithium-ion batteries. *Batteries*, 8(10):148, 2022.

[18] Yanan Wang, Xuebing Han, Languang Lu, Yangquan Chen, and Minggao Ouyang. Sensitivity of fractional-order recurrent neural network with encoded physics-informed battery knowledge. *Fractal and Fractional*, 6(11):640, 2022.

[19] Dingyu Xue and YangQuan Chen. *Scientific Computing with MATLAB.* Chapman and Hall/CRC, 2018.

[20] Donghui Yu, Xiaozhong Liao, and Yong Wang. Modeling and analysis of Caputo–Fabrizio definition-based fractional-order boost converter with inductive loads. *Fractal and Fractional*, 8(2):81, 2024.

[21] Gang Yu, Xianming Ye, Xiaohua Xia, and YangQuan Chen. Digital twin enabled transition towards the smart electric vehicle charging infrastructure: A review. *Sustainable Cities and Society*, page 105479, 2024.

[22] Simeng Zheng, Jiashen Teh, Bader Alharbi, and Ching-Ming Lai. A review of equivalent-circuit model, degradation characteristics and economics of Li-ion battery energy storage system for grid applications. *Journal of Energy Storage*, 101:113908, 2024.

CHAPTER 2

A Review on Fractional-Order Modeling

2.1 INTRODUCTION TO FRACTIONAL-ORDER MODELING

This chapter presents a state-of-the-art survey of the research on fractional-order (FO) modeling with parameter identification and FO estimation methods for state of charge (SOC), state of health (SOH), and remaining usage life (RUL) of lithium-ion batteries (LIBs) mainly in last five years. FO electrochemical models and six different types of FO equivalent circuit models (ECMs) are introduced in detail. Then, the corresponding tuning algorithm for parameters of these FO models are also provided in brief. Moreover, FO estimation methods for SOC are listed and analyzed, including FO observers and FO Kalman filters (FO-KFs). SOH and RUL estimation is another vital aspect for LIBs aging and degradation monitoring; thus FO estimation methods proposed in recent research within five years are all listed. Finally, some suggestions that may be helpful for further research are proposed in conclusion.

Battery is an emerging research aspect due to the increasing energy consumption in current applications and it is the main energy storage device for several types of alternative energy, such as solar, wind, and hydroenergy. As Jeremy Rifkin stated, in current third industrial revolution, there are three pivotal technologies: a communication internet, a renewable energy internet, and a mobility internet, all are connected to the Internet of Things (IoTs) [58]. Basically, batteries are everywhere to construct energy internet and ensure the energy supply for the other two internets in this big data era, so the design and control of batteries are the most concerning aspects for researchers around the world.

According to the charging and storage ways, batteries can be divided into four types: primary battery, secondary battery (rechargeable battery), fuel cell, and reserve battery. Since the 21th century, people are pursuing more sustainable and portable batteries, thus lithium-ion batteries (LIBs) stand out among all kinds of batteries, such as lead-acid battery, Zinc battery, Nickel battery, and hydrogen-oxygen fuel cell. Besides, LIBs have high working voltage, high energy density, relatively low self-discharge, low maintenance, and specific high current to applications, which make LIBs most widely used in applications [67]. However, LIBs have dynamic

DOI: 10.1201/9781003670902-2

non-linearity and aging is always a concern. In Chemistry field, researchers are introducing new and enhanced chemical combinations to improve lithium-ion, while in electrical engineering field, researchers are aiming to learn more about LIBs, then monitoring and controlling LIBs more accurately. This chapter is focused on the electrical aspect, that is, the modeling, and estimation methods for LIBs. Several performance index illustrating information of LIBs can be estimated, such as state of charge (SOC) [15], state of health (SOH) [83], remaining usage life (RUL) [39], and degradation level or ageing level [70]. The well-known ampere-hour (Ah) integral method and the open circuit voltage (OCV) measurement are proposed and commonly used for the estimation of SOC [9, 78, 94]. However, with growth spurt of mobile phones and electric vehicles (EVs), monitoring and management of LIBs in these appliances is faced with higher requirements, which stimulates various prominent modeling and estimation research [8, 36, 66, 85]. Among modeling and estimation research on LIBs, fractional calculus was first applied to present constant phase element (CPE), which started a new fractional research era of LIBs.

Fractional calculus has been initiated for more than 300 years, and started from mathematic definitions, that is, fractional derivative and fractional integral [29, 46]. Further extensions of fractional calculus, such as fractional-order (FO) state space model (SSM) [47], FO PID controller [53], fractional capacitor (CPE) [2], and fractional convection [30, 31], were gradually proposed and applied by electrical engineering researchers in the last 20 years [1, 14, 65]. As to fractional calculus for LIBs research, fractional modeling with identification and estimation of SOC, SOH, RUL are the two main aspects, which are the key monitoring points for further control and management of LIBs. The earliest FO battery research was designed for lead-acid battery about fractional system identification in 2006 [59], when lithium-ion battery has not been widely applied in EVs and mobile devices. After that, some research on fractional modeling of lead-acid battery have been proposed [11, 60]. With LIB gradually replacing lead-acid battery, fractional calculus was first applied to LIB for cell impedance analysis in 2007 [28]. Then fractional impedance analysis, fractional modeling, and fractional estimation methods for LIBs emerged [51, 95, 101], and Figure 2.1 shows the amount and research area distribution of published articles since 2013. The results are searched by "fractional AND battery" and "fractional AND lithium-ion battery" in *Web of Science* and are classified into "LIB" and "Other Battery".

As in Figure 2.1, fractional calculus related research efforts on LIBs have developed rapidly in the past ten years and mainly focuses on engineering, energy fuels, electrochemistry, and automation. To show the states and trends of fractional calculus to LIBs, a literature review on fractional-order research for LIBs over the last five years is presented. Although there are some existing early reviews introducing fractional techniques both on LIBs and supercapacitors (SCs) [4, 21, 77, 88], a specific and detailed analysis only for LIBs is necessary. Hence, this chapter is written to present the novel fractional modeling and estimation methods for LIBs mainly between 2018 and 2023 to make classification of these modeling methods, and then provide some innovative suggestions for future fractional research on LIBs. The chapter is divided into four parts. First, some necessary fundamental knowledge including

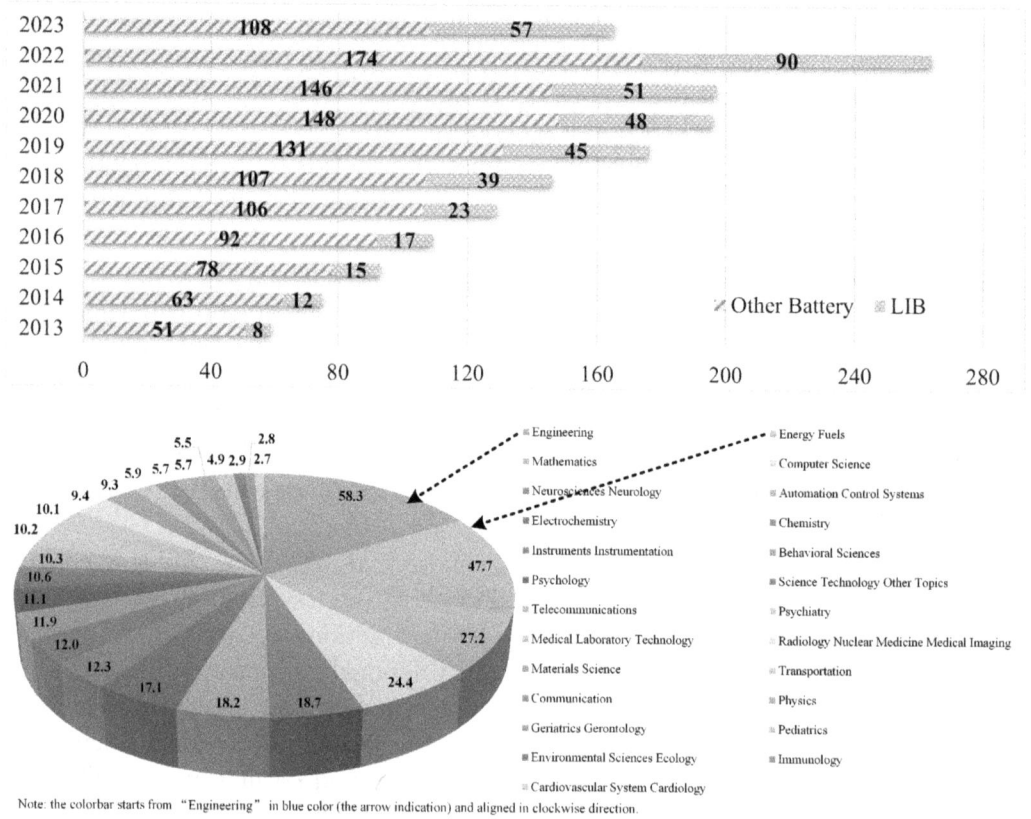

Note: the colorbar starts from "Engineering" in blue color (the arrow indication) and aligned in clockwise direction.

Figure 2.1 Published articles and research areas of "fractional AND battery" since 2013 in *Web of Science*. "LIB" indicates articles related to lithium-ion battery, and "other battery" indicates articles related to others except LIB, and research areas show mainly 25 fields in rank.

SOC, SOH, RUL, and fractional basic definitions is introduced in the second part. Second, fractional modeling and corresponding identification methods are concluded and compared in the third part. Then, all fractional estimation methods of SOC, SOH, and RUL for LIBs in the past five years are provided in the fourth part, including fractional Kalman filter, fractional observer, online estimation, and so on. Finally, the last part provides current challenges and some suggestions for future work about fractional calculus applied to LIBs. It needs to be noted that all the modeling and estimation methods for LIBs mentioned in the following are related to battery cells, rather than battery banks or battery packs in EVs.

2.2 FUNDAMENTAL KNOWLEDGE

This part is designed to provide some basic performance indexes of LIBs, such as SOC, SOH, and RUL. This part also offers brief introduction to fractional calculus and fractional elements, such as fractional Caputo definition, and constant phase element (CPE).

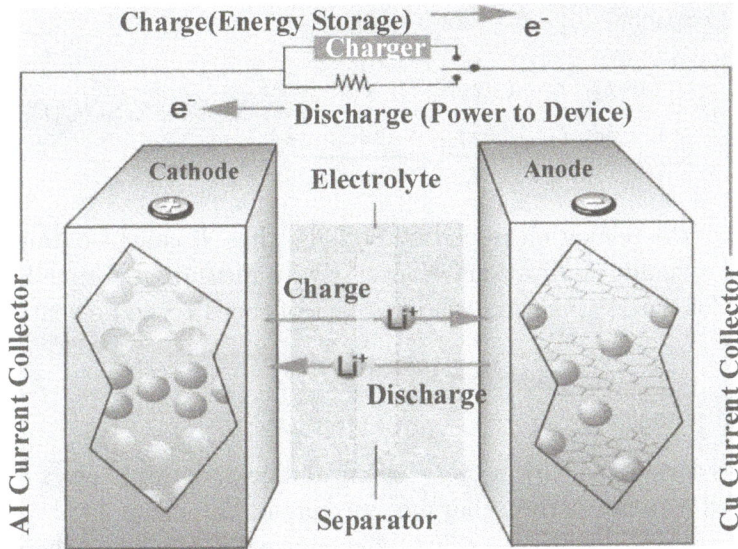

Figure 2.2 Schematic of lithium-ion battery, which consists of four parts: negative electrode (anode), positive electrode (cathode), electrolyte, and separator (modified from [76]).

2.2.1 Performance indexes of LIBs

Figure 2.2 is a schematic of an LIB cell shown in [76], the cell includes four main elements: the positive electrode, negative electrode, electrolyte, and separator. During charging process, lithium ions are transported from cathode into electrolyte and then stored in anode, which builds up a potential difference between the positive and negative electrodes [76]. Discharging is based on the reversed process. The physical and chemical mechanism of an LIB cell can commonly be described by a two-dimensional (2-D) electrochemical model, that is, the Doyle-Fuller-Newman model, which is rarely used in real-time battery management system (BMS) due to the prohibitive computation [17]. Hence, single particle model (SPM) is derived by neglecting the electrolyte dynamics, so that the 2-D electrochemical model is simplified to one spatial dimension [57]. Due to the heterogeneity in the chemistry process, LIBs would show up some stochastic behaviors, like diffusion effect, leak current, and self-discharge, which can affect LIBs working states. To illustrate transient states of LIBs, several performance indexes have been defined. Three key indexes are introduced in this part, that is, state of charge (SOC), state of health (SOH), and remaining usage life (RUL). Furthermore, fractional calculus is also applied to build the governing equations of an SPM for these reactions, specifically for the diffusion phenomena [50], and the details will be introduced in Section 2.3.1.

2.2.1.1 State of charge

SOC illustrates the remaining amount of available charge $Q(t)$ in an LIB, and cannot be directly measured. SOC can be expressed as the remaining percentage of a reference capacity Q_{ref} as follows [16],

$$SOC = \frac{Q(t)}{Q_{ref}} = \underbrace{\frac{Q(t_0)}{Q_{ref}}}_{SOC_0} + \underbrace{\frac{\int_{t_0}^{t} I(\tau)d\tau}{Q_{ref}}}_{\Delta SOC(t)} = SOC_0 + \Delta SOC(t). \tag{2.1}$$

In practice, the relaxation period is too long; thus, it cannot obtain SOC_0 when LIB works in dynamic applications. Hence, how to measure real-time SOC or design an estimator for SOC estimation is a key point of the BMS to prevent overcharge or overdischarge.

2.2.1.2 State of health

SOH is always considered with battery ageing and degradation process of LIBs, which can be affected by temperature, charging current, and discharge level [87]. SOH does not have the specific definition as SOC, but is usually defined as the ratio between remaining capacity and initial nominal capacity, as shown in the following [77],

$$SOH = \frac{C_p}{C_0}, \tag{2.2}$$

where C_p represents the available capacity and C_0 represents the rated capacity specified per design and measured under the aforementioned conditions prior to operation. From equation (2.2), we can see that SOH decreases with fading capacity and rising ohmic resistance due to loss of active electrodes, solid electrolyte interphase (SEI), and irreversible lithium reactions [6].

There are four significant points for the aging of battery indicated by SOH as listed below [48]:

1. FL-BOL: First life-beginning of life, serves as the reference for all the internal cell parameters monitored ($SOH = 100\%$).

2. FL-EOL: First life-end of life, represents the moment when the batteries are retired from the automotive use ($SOH = 80\%$) [71].

3. SL-BOL: Second life-beginning of life, represents the moment when the batteries are reimplemented on a second life applications.

4. SL-EOL: Second life-end of life, reproduces the moment of second life battery retirement.

2.2.1.3 Remaining usage life

RUL is the corresponding definition of SOH in the aspect of battery aging cycles amount. As mentioned in Section 2.2.1.2, LIB needs to be replaced when reaching the FL-EOL point. Hence, there is a certain limited LIB cycles amount before the cycle life threshold, and RUL means the remaining cycles before this threshold. The RUL estimation is always companied with SOH estimation, and the estimation methods vary from internal resistance measurement of physics-based model to data-driven prognostics [38]. The detailed introduction will be provided in the Section 2.4 of this chapter.

2.2.2 Fractional calculus and fractional elements

2.2.2.1 Fractional calculus

Fractional calculus is based on FO integrals and FO derivatives, also called non-integer-order integrals and derivatives [56], which can also be divided into left and right ones [32]. Since FO derivatives are applied to LIBs modeling and estimations rather than FO integrals, only left fractional derivatives are presented in the following. Basic fractional definitions commonly include Riemann-Liouville (R-L) definition, Grünwald-Letnikov (G-L) definition and Caputo definition [5], which may not be equivalent to each other. First, the three types of fractional derivatives are provided in the following. The R-L definition is expressed as [32]

$$^{RL}D_t^\alpha f(t) = \frac{1}{\Gamma(n-\alpha)}\frac{d^n}{dt^n}\int_a^t \frac{f(\tau)}{(t-\tau)^{\alpha-n+1}}d\tau, t > a, \qquad (2.3)$$

where $f(t)$ is an arbitrary integrable function in $[a, b]$, $\alpha \in (n-1, n)$, $^{RL}D_t^\alpha$ represents the R-L type derivative operator, and $\Gamma(\cdot)$ is the Gamma function. The G-L definition is expressed as [91]

$$^{GL}D_t^\alpha f(t) = \lim_{h\to 0} h^{-\alpha} \sum_{j=0}^{[\frac{t-t_0}{h}]} (-1)^j \begin{pmatrix} \alpha \\ j \end{pmatrix} f(t-jh), \qquad (2.4)$$

where $^{GL}D_t^\alpha$ represents the G-L type derivative operator, $[\frac{t-t_0}{h}]$ is the approximate recurrence term for integer part, and $\begin{pmatrix} \alpha \\ j \end{pmatrix} = \frac{\alpha!}{j!(\alpha-j)!}$ represents the coefficient of the recursive function. Caputo definition is another widely used one in engineering and control field due to the same initial conditions with integer-order (IO) derivative. The Caputo definition is expressed as [91]

$$^{C}D_t^\alpha f(t) = \frac{1}{\Gamma(n-\alpha)}\int_a^t \frac{f^{(n)}(\tau)}{(t-\tau)^{\alpha-n+1}}d\tau, t > a, \qquad (2.5)$$

where $^{C}D_t^\alpha$ represents the Caputo type derivative operator. According to [32], if $f(t)$ is suitably smooth, i.e. $f \in C^n[a, b]$, then the R-L derivative and the G-L derivative are equivalent, that is, $^{RL}D_t^\alpha f(t) = {^{GL}D_t^\alpha} f(t)$; the R-L derivative and Caputo derivative have the following equation

$$^{RL}D_t^\alpha f(t) = {^{C}D_t^\alpha} f(t) + \sum_{k=0}^{n-1} \frac{f^{(k)}(a)(t-a)^{k-\alpha}}{\Gamma(k+1-\alpha)}, \qquad (2.6)$$

where $n - 1 < \alpha < n$, $f \in C^{n-1}[a, t]$ and $f^{(n)}$ is integrable on $[a, t]$. However, for LIB research, G-L definition is easy to be discretized in time domain; thus, it is most commonly applied to fractional time-domain models or fractional estimators for LIBs. Besides, the Laplace transform of Caputo definition under zero initial conditions is $\mathcal{L}_0^C D_t^\alpha f(t) = s^\alpha F(s)$, which is suitable for fractional research in frequency domain. Hence, Caputo definition and G-L definition are more widely used in LIB modeling and estimations. The reason is that the initial conditions of Riemann-Liouville definition have more complicated forms than that of Caputo definition as shown in equation (2.6). Besides, the computational load of FO derivatives are heavier than integer-order ones, so some numerical methods have been designed in [29, 32].

2.2.2.2 Fractional elements

Besides basic fractional calculus definitions, "fractor" is another essential element for modeling and estimation of LIBs. It is well-known that capacitors and inductors are not ideal ones in practical system, and the relationship between voltage and current is not just first-order derivative or integral, especially in low frequency or high frequency. Hence, the term "fractor" arose following the successful synthesis of an FO capacitor or an inductor, and the transfer function of fractor is given by [2],

$$Z_F(s) = \frac{1}{F s^\alpha},\qquad(2.7)$$

where F is the impedance of the fractor, named as "fractance". Fractor is also called constant phase element (CPE) , and FO capacitor and fractional-order inductor are the two types of fractor in nature, respectively. Equation (2.8) shows the voltage-current relationship of a typical FO capacitor [91]:

$$\begin{cases} i(t) = C_w \frac{d^\alpha u(t)}{dt^\alpha}, & 0 < \alpha < 1, \ t \geq 0 \\ \frac{U(s)}{I(s)} = \frac{1}{C_w s^\alpha}, \end{cases}\qquad(2.8)$$

where C_w is a constant related to the capacitance, and α represents the order of the FO derivative in Caputo definition. From equations (2.7) and (2.8), CPE was first proposed to replace the IO capacitors inside LIB models and explain the low-frequency dynamics of LIB. A similar fractional element is called Warburg element, which is a 0.5th order CPE in a typical Randles model for LIB [68]. In the past five years, Warburg element turns to be any fractional order instead of 1\2, same as CPE. Hence, CPE is the basis of FO electrical circuits for LIBs, which will be further introduced and discussed in the following sections.

2.3 FRACTIONAL-ORDER MODEL FOR BATTERY

In BMS, it is always necessary to build a model with parameter identification before other estimations, monitoring, and charge or discharge control. Plenty of research has been published in the modeling and parameter identification aspects of LIBs [40, 80, 92]. As fractional calculus is extended from integer calculus [32], FO modeling and corresponding identification are also vital extensions. In an early survey of FO techniques applied to LIBs, lead-acid batteries, and SCs [101], four kinds of typical FO circuit models for LIBs have been offered. In this section, a more complete sets of FO models are presented, including some new research published in the past two years. Then the corresponding parameter identification methods for these LIB models are also provided in Section 2.3.3.

FO modeling of LIBs mainly can be divided into two aspects, that is, electrochemical model (equations) and ECM. The main difference between electrochemical model and ECM is the presentation elements for battery mechanism and inner reactions. ECM applies electric components to build an equivalent circuit for battery reactions, while electrochemical model applies electrochemical differential equations to reflect the inner reactions of LIBs. Thermal and aging models for LIBs are also investigated in recently years due to their important roles in the kinetics of charge transfer process and side reactions introduced in Section 2.2.1 and shown in Figure 2.2 [50].

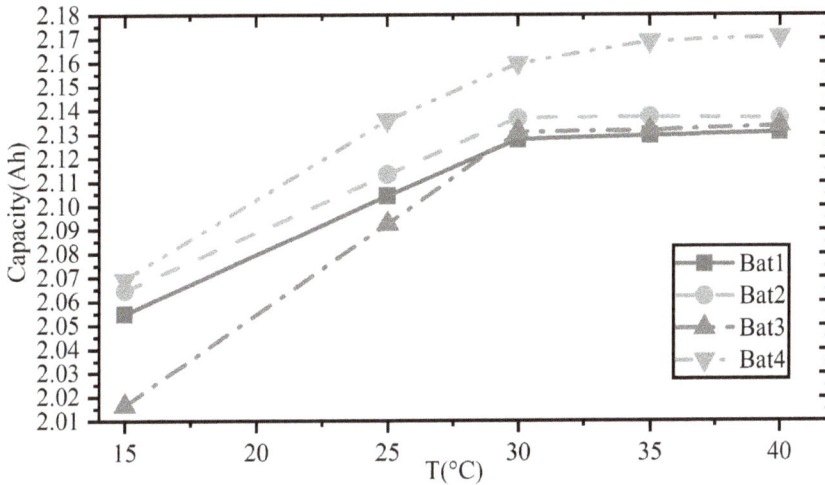

Figure 2.3 The available capacity of four 18650 LIBs (rated capacity = 2Ah, named as Bat1, Bat2, Bat3, and Bat4) in five temperatures (15°C, 25°C, 30°C, 35°C, and 40°C).

Figure 2.3 presents the available capacity of four 18650 LIBs (rated capacity = 2Ah) in five temperatures, that is, 15°C, 25°C, 30°C, 35°C, and 40°C. It illustrates that the available capacity of a battery cell varies much in different temperatures, especially in low temperatures. The low temperature situation may happen in EVs' battery pack during winter. Hence, the model parameters and performance indexes estimations (SOC, SOH, RUL) would be influenced by the temperatures, so that FO thermal models is another necessary investigation direction for LIBs. While integer-order thermal and ageing models are also two vital aspects for modeling of LIBs, the extended FO ones are still very few and worthy of further investigation. Moreover, the existing research on FO thermal and aging models are considered together with electrochemical model [50], and this chapter is inclined to more electrical engineering aspects, so the thermal and aging models are just briefly discussed together with the FO electrochemical model in Section 2.3.1.

2.3.1 Fractional-order electrochemical model

This type of FO model was first proposed by Sabatier et al. in [63]. The FO model is converted from a typical electrochemical model, called single particle model (SPM), which was built on the basis of the electrochemical reactions inside a lithium-ion cell. A typical SPM generally includes four partial differential equations (PDEs) describing four key variables of the electrode and electrolyte, that is, lithium concentration c_{se} in the spherical particle by diffusion law, lithium concentration c_e in electrolyte, charge conservation in electrode (electrode potential ϕ_s) by Ohm's law, and charge conservation in electrolyte (electrolyte potential ϕ_e), respectively [64]. Then all of the four differential equations are linked by the Bulter-Volmer equation. As mentioned in Section 2.2.1, an SPM is derived by neglecting the electrolyte dynamics and treating each electrode as a spherical particle that stores Li^+ as shown in Figure 2.4. From [50], FO electrochemical modeling for LIBs is based on the solution of Fick's first law of diffusion, and lithium ions concentration gradient in the particle can be described by the following:

$$\frac{\partial c_{se}}{\partial t} = \frac{D_{se}}{r^2}\frac{\partial}{\partial r}(r^2\frac{\partial c_{se}}{\partial r}) \left\{ \begin{array}{l} \frac{\partial c_{se}}{\partial t}\Big|_{r=0} = 0 \\ D_{se}\frac{\partial c_{se}}{\partial r}\Big|_{r=R_s} = -\frac{j^{Li}_{mean}}{a_s F} \end{array} \right. \tag{2.9}$$

where c_{se} is the lithium ion concentration, D_{se} is the diffusion coefficient, r is the radius of the sphere, and j^{Li}_{mean} is the average current density.

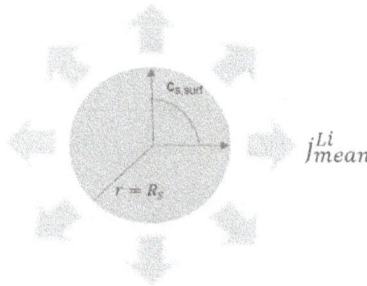

Figure 2.4 Single particle model with concentration gradient through the sphere (modified from [50]).

The analytical solution of equation (2.9) is a transfer function, linking the mean value of the lithium current density in the electrode $J^{Li}_{mean}(s)$ to lithium concentration c_{se}. Based on the traditional SPM, Sabatier et al. have found that this transfer function can be approximated using the fractional transfer function [61]

$$H_{csi,e}(s) = \frac{c_{se}(s)}{J^{Li}_{mean}(s)} = \frac{K_{1i}(1 + \frac{s}{\omega_{csei}})^{0.5}}{s}. \tag{2.10}$$

Based on equation (2.10) and assuming that the electrolyte potential is constant, an SPM can be simplified to a single-electrode model shown in Figure 2.5, if removing the negative electrode contribution [16, 62]. In this way, the FO electrochemical

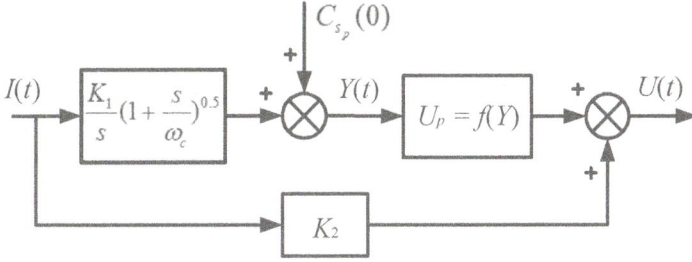

Figure 2.5 Single-electrode model in frequency domain (modified from [62]).

model has a concise structure without using large number of model parameters, but still holds the accuracy to reflect electrochemical dynamics.

Sabatier et al. have published a new research related with this fractional-order electrochemical model in 2018 [50], which combined with an efficient simple thermal model and an aging model designed for generation of fast charging algorithms. Like the FO electrochemical model, thermal and aging models are presented in partial differential equation (PDE) forms. For example, the efficient thermal model is mainly based on a heat transfer equation as

$$mC_{\mathrm{p}}\frac{dT(t)}{dt} = Q_{\mathrm{gen}}(t) - Q_{\mathrm{loss}}(t). \tag{2.11}$$

where m is the mass of the cell, C_{p} is the specific heat capacity, Q_{gen}, and Q_{loss} are the generated heat and convective heat with the environment, respectively. As to the aging model, the structure is driven from the degradation caused by formation of solid electrolyte interphase (SEI) layer growth on the anode shown in Figure 2.2, and the detail PDEs in detail can also be found in [50]. Similar fractional transfer function approximation has been applied to electrochemical equations in [34], and a simplified state space model of battery terminal voltage and load current was proposed and transferred into discrete form for further research. Since this type of FO model for LIB is a simplified one for SPM, it may also be useful to other kinds of enhanced SPM, like SPM with electrolyte dynamics (SPMe), SPM with electrolyte and thermal dynamics (SPMeT) [55]. Moreover, the FO thermal and aging models still have lots to be explored and are worthy of further research. While the fractional calculus applied in the electrochemical models can better illustrate the dynamic reactions or thermal influences for LIBs, the FO PDEs of the FO electrochemical models remain computationally expensive for real-time BMS, which is also an aspect to be improved.

2.3.2 Fractional-order equivalent circuit model

As to the FO ECM, it is another electrical engineering way to analyze LIBs dynamics, while temperature and aging are converted into parametric functions or SOC, SOH estimations [102]. Almost all ECMs structures were proposed according to the electrochemical impedance spectrum (EIS) of LIBs as shown in Figure 2.6, because the EIS reveals the electrochemical dynamics in frequency-domain and varies with

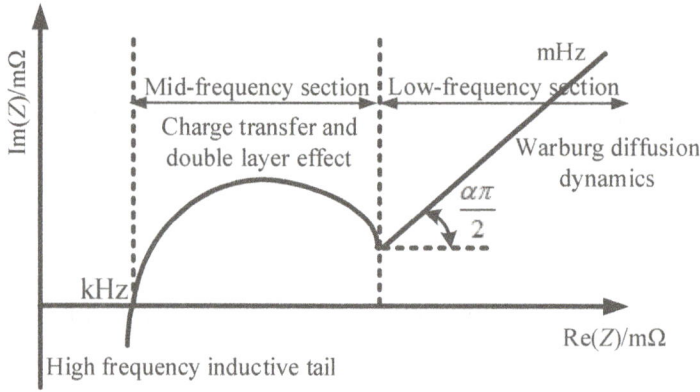

Figure 2.6 A typical schematic diagram of EIS of LIB.

SOC, SOH, RUL, and temperature of LIBs. So various kinds of ECMs were proposed to explain the three main parts of the EIS, that is, high-frequency inductive tail, mid-frequency reaction, and low-frequency Warburg diffusion dynamics. Here four forms of ECMs are presented in Figure 2.7, and each form can be separated into certain types of ECMs in the following sections.

2.3.2.1 FO Thevenin model $(n = 1)$

If n in Figure 2.7(a) is 1, ECM in Figure 2.7(a) becomes an FO Thevenin model, which is also called 1-RC model in traditional integer-order model. The state space model (SSM) of FO Thevenin model in Caputo definition was proposed and approximated into a discrete system in [102]. Since FO Thevenin model is already a simplified one and the approximation accuracy requires heavy computation, a data-based FO Thevenin model was built in continuous-time form [26]. From Figure 2.7(a) $(n = 1)$, the FO Thevenin model is proved to be a simpler fractional ECM for LIB on the basis of the EIS and hybrid pulse power characteristic (HPPC) test [75]. As it only considers high-frequency and mid-frequency reactions, and ignores Warburg element of diffusion effects [52]. Hence, FO Thevenin model is seldom applied in the last five years in fractional modeling. In comparison, FO Partnership for a New Generation of Vehicles (PNGV) and FO "Randles" model are the more widely used models.

2.3.2.2 FO PNGV and Randles model

Figure 2.7(b) and Figure 2.7(c) are the FO Randles model and FO PNGV model, respectively. Both of them are systems with two fractional orders. Actually, traditional Randles model is already an FO system, because the Warburg element reflecting diffusion dynamics was always considered as 0.5th order in the previous research. Since the diffusion effect varies due to different temperature, SOC, aging level, the first step is changing the 0.5th order Warburg element into an arbitrary fractional order element, that is, another CPE [86]. Then, Wang et al. have found that the curve in mid-frequency shows a semi-ellipse rather than semi-circle due to capacitance

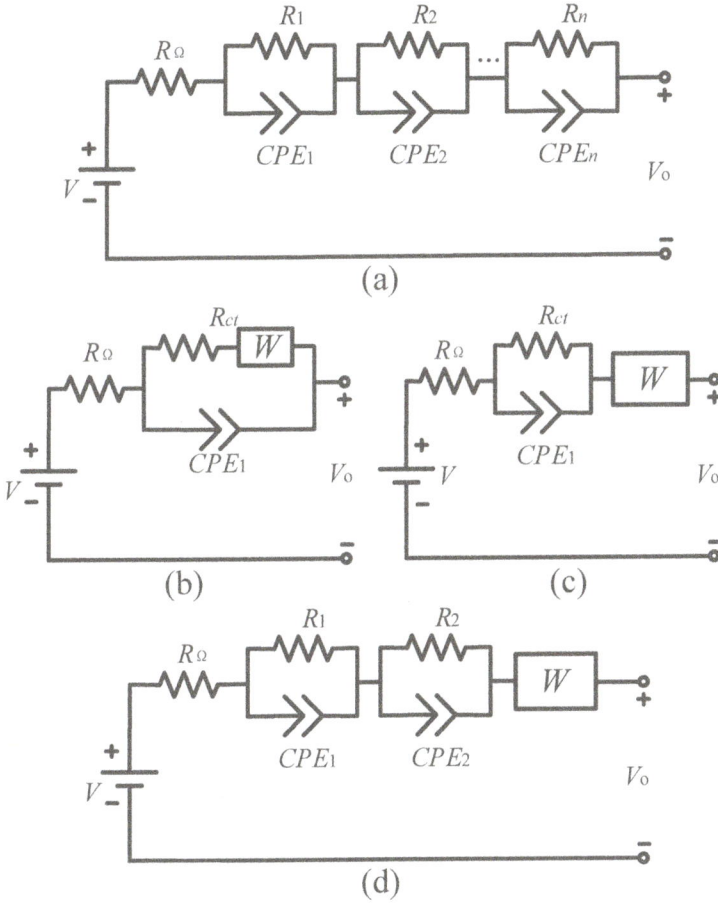

Figure 2.7 Four forms of ECMs for LIBs., (a) n-RC^α Model, (b) FO Randles model, (c) FO PNGV model, (d) three-orders FO model ($CPE_i, i = 1, 2, ..., n$ and W are the fractional elements introduced in Section 2.2.2.2).

dispersion, so the ideal capacitor reflecting double layer effect was replaced by a CPE in [72], constructing the FO Randles model in Figure 2.7(b).

On the other side, FO PNGV model is also a quite popular FO model for LIBs, and the corresponding state-space model (SSM) is denoted by the followings [37]

$$
\begin{bmatrix} \frac{d^m U_W}{dt^m} \\ \frac{d^n U_{cpe}}{dt^n} \end{bmatrix} = \begin{bmatrix} 0 & 0 \\ 0 & -\frac{1}{R_{ct}C_{cpe}} \end{bmatrix} \begin{bmatrix} U_W \\ U_{cpe} \end{bmatrix} + \begin{bmatrix} \frac{1}{C_W} \\ \frac{1}{C_{cpe}} \end{bmatrix} I,
$$
$$
U_0 = \begin{bmatrix} 1 & 1 \end{bmatrix} \begin{bmatrix} U_W \\ U_{cpe} \end{bmatrix} + R_\Omega I + U,
$$
(2.12)

where U_W and U_{cpe} mean the voltages of Warburg element and CPE_1; C_W and C_{cpe} mean the capacitance of Warburg element and CPE_1; m and n mean the fractional order of Warburg element and CPE_1, respectively; I is the current of the FO PNGV model.

It can be known from equation (2.8) that FO PNGV model is an incommensurate order SSM, while some research would make $m = n$ to simplified into a commensurate order one. Xiao et al. have made a comparison among FO PNGV, FO Thevenin, IO PNGV, and IO Thevenin models [79]. The results prove that FO PNGV model can describe the OCV variation and low-frequency dynamics of LIB, and better capture the dynamic performance than the other three kinds of models. Moreover, it is interesting that a simplified FO Randles model analyzed in frequency domain [54] may have some connections with FO PNGV model. The authors of [54] separated the charge transfer process with diffusion dynamics [13], which turns out to be the structure of FO PNGV model. The Nernst diffusion phenomenon is first involved in [54], then simulated by FO integrator, which turns out to be a 0.5 order integrator. Hence, it seems that the simplified Randles models in [54] is a specific FO PNGV model.

2.3.2.3 2-RC^α model ($n = 2$)

If in Figure 2.7(a) $n = 2$, ECM in Figure 2.7(a) becomes an FO system with two fractional orders. This type of FO model was proposed because the low-frequency part of EIS is proven to be a part of a depressed semicircle with a large diameter rather than a straight line, by certain dynamic tests, such as hybrid pulse power characteristic (HPPC) tests [84, 96]. Hence, the Warburg element is represented by the parallel combination of a CPE and a resistance, which is also called "ZARC" element [7]. The structure has little difference with FO PNGV model, but would bring more computation burden for further investigation, especially when the two fractional orders are incommensurate. However, more research used this 2-RC^α models recently [41, 73].

2.3.2.4 High-order FO model ($n \geq 3$)

For higher modeling accuracy to fit EIS, some researchers have proposed extended high-order FO models for LIB. Hu et al. have improved the 2-RC integer-order model by adding low-frequency component and replacing ideal capacitor with CPE [22, 33], which results in a high-order FO model with three fractional orders α, β, γ as in Figure 2.7(d). Another type of high-order FO model with three fractional orders is based on the FO Randles model [98]. Similar to [84], a "ZARC" element was added to describe the high-frequency part of EIS, which intersects with the mid-frequency part. Moreover, Jacob et al. have proposed a general FO battery EIS model with the structure in Figure 2.7(a), which has n CPEs with n fractional orders [25]. The number of parallel tanks depends on the required accuracy, and the parallel resistance can be neglected to build a Warburg term, so that this general EIS model with n fractional orders holds high flexibility.

2.3.2.5 Variable and specific FO models

Considering the fractional orders may change with several working factors (time, temperature, aging), some researchers have proposed variable FO models for LIBs. Hu et al. constructed a monotonous relationship between SOC and the fractional order, which was considered reflecting the fractal morphology of charge distribution [43]. Then, Lu et al. also applied the fractional order as an indicator for electrode aging [42]. In this way, variable FO model provides a rapid method to estimate SOC and evaluate aging level. However, the real-time identification for fractional orders of the variable FO model is a tough task work in some practical application which may require online adaptive algorithm [7].

Despite those models cited above, some specific FO models can also provide novel explanation for LIBs, like the three types of FO impedance models involving bounded diffusion with three kinds of particle geometry [18]. Also Zhang et al. combined kinetic battery model (KiBaM) with ECM to build a hybrid FO model [89]. Considering the linear requirement of EIS test, Xiong et al. replaced charge transfer resistance by Bulter-Volmer (BV) equation and ohmic resistance by a piecewise quadratic function of current, resulting in a BV-FOM [82].

2.3.3 Parameters and fractional-orders identification

Before all the FO models applied further to estimation or control, parameter identification is the first step to ensure the accuracy of FO models. Compared to integer-order models, the added fractional orders increase the identification difficulty for the researchers, so they are also searching effective tuning methods. Here state-of-the-art tuning methods are introduced briefly, and some suggestions are offered.

In earlier period of FO modeling, the Thevenin model was often applied because the corresponding identification methods were simpler, e.g. step response curve [61], algebraic calculation [90, 93], least squares (LS) method [27], and gradient method [13]. With more complicated FO modeling including two or three fractional orders, optimization algorithm and adaptive observer were increasingly designed [3, 12]. In the observer aspect, a Kreisselmeier-type adaptive observer has been proposed for FO Randles model [68, 69]. As for the optimization part, genetic algorithm (GA) and particle swarm optimization (PSO) are the two most applied algorithms for FO models among the large amount of optimal algorithm [72, 98].

In the past three years, people tend to design more enhanced algorithms for $2\text{-}RC^\alpha$ and high-order FO models, like LS-GA (combination of the LS methods and GA) [84], and mixed-swarm-based cooperative particle swarm optimization (MCPSO) [22]. These long-name algorithms have complicated procedures and high cost, which may not be available in practical battery working situations. Investigating online algorithms like LS improved adaptive method in [75], or automatic updating methods, like the automatic updating of parameters values at different aging stages in [35] are more desirable. Moreover, the analysis of specific algorithm and influencing factors, such as Bayesian inference [25] and the historical data dynamics [24], are also very necessary for other researchers to refer in their new FO modeling for LIBs.

2.4 FRACTIONAL-ORDER ESTIMATION

Either fractional modeling or parameters identification aims to build an accurate model of LIB for further monitoring or control. For battery monitoring, SOC, SOH, and RUL are three main performance indexes to indicate dynamic working states of LIBs. Based on different kinds of FO models, FO estimation methods for SOC, SOH, and RUL are also proposed in last five years, which will be presented in this section.

2.4.1 SOC estimation methods

The traditional SOC estimation methods for LIBs generally include four aspects: basic methods (Ah and OCV methods), model-based observers, model-based Kalman filter (KF) series, and machine learning (ML). From equation (2.1), it is obvious that the Ah method is the simplest direct way to estimate SOC, its definition is presented as follows [10]:

$$SOC(t) = SOC_0 + \frac{\int_0^t \eta i(t)}{Q_N} dt, \tag{2.13}$$

where SOC_0 is the initial SOC value, Q_N is the rated battery capacity, and η is the charge-discharge efficiency. Ah method is simple but it depends on current measurement accuracy and has accumulated error. Thus, the monotonous relationship of OCV to SOC was proposed, however, OCV needs to be measured after long time rest of LIBs, which is not possible in some practical working situations. Hence, model-based methods are the main focus in recent research, such as sliding mode observer (SMO), Luenberger observer, and various types of KFs, while Ah and OCV methods are usually applied as SOC reference in current research.

By analyzing the published articles of SOC FO estimation for LIBs from 2014 to 2019, these methods can mainly be divided into four types: SMO and Luenberger observers, other observer, FO Kalman filter series, and special estimator, as shown in Figure 2.8(b). It is obvious that FO-KF series methods were investigated and proposed mostly. To better illustrate the state-of-art research distribution, four main aspects of traditional methods are also collected from *Web of Science* by searching

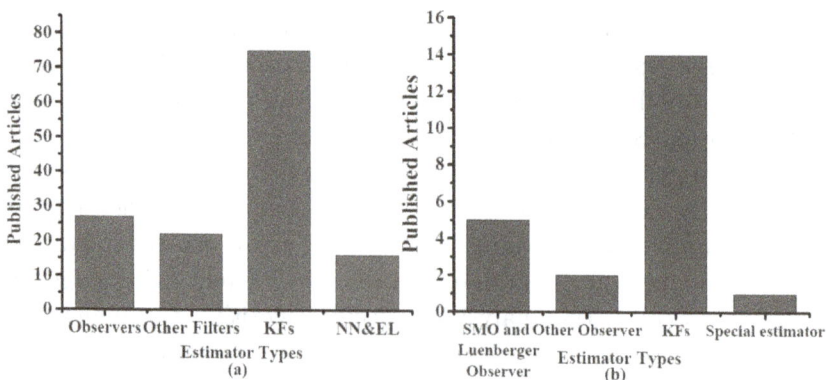

Figure 2.8 Distribution of published articles about SOC estimation methods for LIBs., (a) Four main aspects of the traditional estimation methods: observers, other filters, KFs, and neural network (NN) & extreme learning (EL); (b) four main FO estimation methods: SMO and Luenberger observer, other observer, KFs, and special estimator.

"lithium-ion battery", "state of charge", and corresponding key words, as shown in Figure 2.8(a). Since various kinds of FO ECMs listed in Section 2.3.2 have been applied to LIBs modeling, the traditional model-based estimation methods can be directly transferred to FO estimation methods. However, other fractional filters and fractional neural network were already proposed but still not applied, which may be the new direction of FO methods. In the following, we respectively present the methods shown in Figure 2.8(b).

2.4.1.1   FO Luenberger observer and SMO

The FO SMO was firstly proposed for SOC estimation of LIBs based on a 2-RC^α FO Model with commensurate fractional order α as shown in Figure 2.7(a)($n = 2$) [95]. Zhong et al. derived the differential equations for the 2-RC^α FO ECM under uncertainty $\delta_i, i = 1, 2, 3, C_e, C_d$ disturbance caused by non-linear dynamics. Then, considering the fractional dynamic system [96],

$$D^\sigma X = f(X, I_{in}) + BW(t), \quad \sigma \in (0, 1], \tag{2.14}$$

where X denotes state variable, $f(X, I_{in})$ is the system function, and B and I_{in} are the constant matrix and the input vector, respectively. The proposed SMO is that

$$D^\sigma \hat{X} = f(\hat{X}, I_{in}) + L_i \mathrm{sgn}(X - \hat{X}), \tag{2.15}$$

where $\hat{X}$ is the estimation for X and L_i is the SMO gain. Thus, the SMO in equation (2.15) was applied to the differential equation including SOC variable $S(t)$, to ensure the estimation error to approach zero. Zhong et al. have reduced SMO chattering by adjusting the SMO gain L_i, resulting an adaptive FO SMO [96], and also estimated the polarization voltage at the same time in a later article [97]. Similarly, a Luenberger observer was designed for a 2-RC^α FO Model with the same structure, but with incommensurate fractional orders α_1 and α_2 [73]. The structure of the Luenberger observer is also similar to equation (2.15) but without function sgn($\cdot$). Based on SMO and Luenberger observer, Zou et al. designed a non-linear FO estimator for LIBs with FO PNGV model [100], and the non-linear FO estimator has the form as follows:

$$D^\alpha \hat{x} = A\hat{x} + Bu + H_0(\hat{x}, u)$$
$$+ L_l(y - \hat{y}) + L_s \mathrm{sgn}(y - \hat{y}), \tag{2.16}$$
$$\hat{y} = C\hat{x} + Du + f(\hat{x}).$$

The detailed parameters and variables definitions in equation (2.16) can be referred to [100]. From equation (2.16), the non-linear FO estimator is actually the combination of SMO and Luenberger observer, and the original SSM of the FO PNGV model is

$$D^\alpha x(t) = Ax(t) + Bu(t) + H_0(x, u),$$
$$y(t) = Cx(t) + Du(t) + f(x_1(t)), \tag{2.17}$$

where $u(t) = [SOC(t), V_{cpe}(t), V_W(t)]^T$, $y(t)$ is the output voltage V_o, and $f(x_1(t))$ is a non-linear function related to SOC. It needs to be noted that equation (2.17) is a more useful SSM expression for FO PNGV model instead of SSM in equation (2.12), because SOC is inherently included in the state variables.

2.4.1.2 FO Kalman filters

As the most commonly used kind of filters, FO-KFs for SOC estimation of LIBs includes several types of KFs, such as KF [45], extended KF (EKF) [37], unscented KF (UKF) [10, 74], cubature KF (CKF) [44], and dual KFs. All kinds of KFs are able to eliminate estimation error depending on the five basic equations including state time update, state measurement update, gain matrix update, time update of error covariance, and measurement update of error covariance [79]. While fractional-order KFs are still based on the five basic equations, the state update equations are in discrete forms with fractional differential. Since KFs are working in an iterative process with discrete forms, the original estimated FO continuous-time model needs to be discretized, which uses G-L definition in all research. Table 2.1 lists all types of FO-KFs applied to SOC estimation in last five years, and the corresponding comments and analysis are provided in the following.

Table 2.1 All types of FO-KFs for SOC estimation in the last five years

No.	FOKFs Types	References	Estimated FO-ECM
1	FO-KF	[41]	2-RC^{α_1,α_2} model
2	FO-EKF	[49]	2-$RC^{\alpha,\beta}$ model
3	FO-AEKF	[99]	2-$RC^{m,n}$ model
4	UKF	[52]	FO PNGV model
		[82]	BV-FO Thevenin model
5	Dual FO-KF	[33]	three-orders FO model
6	Dual FO-EKF	[23]	2-$RC^{\alpha,\beta}$ model
7	Dual FO-UKF	[7]	2-RC^{α_d,α_e} model
8	FO-CKF	[44]	FO Thevenin model

It needs to be noted that not all references of a type of FO-KF are listed in Table 2.1. Instead, a typical reference is chosen to be analyzed here, because the other references with same type are similar and listed in the references of this chapter. The following comments and analysis are provided:

1. FO-AEKF (FO Adaptive EKF) in [99] is able to change the process noise covariance matrices Q and measurement noise covariance matrices R with the estimation process, which is very intelligent and ensures high estimation accuracy in practical dynamic working situations.

2. Dual FO-UKF in [7] is for SOC and fractional-order estimation, dual FO-EKF in [23] is very efficient for SOC and SOH estimation at the same time, and dual FO-KF in [33] is for SOC estimation and parameter identification. The dual structure can avoid applying other algorithm, reduce the system complexity, and improve the estimation efficiency.

3. Dual FO-KF in [33] is also an adaptive KF (AKF) with the real-time parameter update, which is suitable for online SOC estimation.

4. In all references, FO-KFs was compared with integer-order KF, EKF, or UKF, resulting in faster convergence speed and higher accuracy. The reason to choose FO-KF, FO-EKF, or FO-UKF needs to be further investigated, like for non-linear dynamics, and a comparison between these FO-KFs may be necessary.

5. Most of FO-KFs are based on FO ECMs, especially 2-$RC^{\alpha,\beta}$ model, so the design process may be similar. FO-KFs based on other kinds of FO models for LIBs can be discussed further, like the improve model by BV equation in [82], and the FO-AEKF designed in [34].

2.4.1.3 Other FO observers and special estimators

Another observer based on FO electrochemical model was proposed by Sabatier et al. in [16] and [62]. It is the same research group introduced in the electrochemical model as mentioned in Section 2.3.1, and the electrochemical model is actually an FO model with 0.5th order. The SOC estimation was implemented by employing two error injection schemes, that is, "input current feedback" and "SOC feedback" [16]. The structure of SOC feedback observer is much like a Luenberger observer. Here a special SOC estimation method is presented, that is, a rapid SOC estimation by fractional order [43], which has proposed that the fractional order can indicate SOC. The basic principle is finding the relationship between SOC and fractional order, and building a look-up table for the iterative estimating process to search.

From SOC estimation methods presented above, it still has many other aspects to investigate. From Figure 2.8, the model-based methods dependent on the FO ECMs is the mainstream of current SOC FO estimation research, however, machine learning (ML) is another developing aspect and has much to do in future, such as fractional-order neural network and FO version of extreme learning (EL) for SOC estimation of LIBs.

2.4.2 SOH and RUL estimation methods

SOH and RUL are connected with each other, so they are discussed together in this part. Since the degradation of LIBs are non-linear dynamics, SOH and RUL estimation cannot be considered as linear measurements. Hence, fractional calculus provides a novel way to estimate SOH and RUL during the aging process of LIBs. However, fractional-order SOH and RUL estimation methods proposed in last five years are not as many as those for SOC. One reason is that SOH and RUL do not have a unique definition that is easy to be quantified and the second reason is that battery aging problems is difficult to be described by fractional calculus, like end of life (EOL) and second-life reuse problems . From Section 2.2.1.2, SOH can be estimated by remaining capacity, resistance, and related to lifetime, degradation level, RUL of LIB. Here an SOH index is shown as [19]

$$SOH_R = \frac{R_{EOL} - R_{current}}{R_{EOL} - R_{init}}, \tag{2.18}$$

where R_{EOL}, $R_{current}$, and R_{init} are resistance values at EOL, current status, and fresh status, respectively. Hence, the resistance of FO model can be estimated to calculate SOH, which may be related with EIS measurement of LIB. Similarly, RUL, degradation, and aging level can also be investigated by the resistance or even the fractional order [42, 66]. Table 2.2 lists all published articles related with FO estimation methods for SOH, RUL, degradation, lifetime, and battery aging. Some brief explanations and analysis are listed in the following.

Table 2.2 SOH and RUL estimation methods (2014–2019)

No.	References	FO-ECMs	Methods
1	[23]	2-$RC^{\alpha,\beta}$ model	Dual FO-EKF
2	[71]	2-$RC^{\alpha,\beta}$ model	Incremental capacity analysis
3	[20]	FO PNGV model	Overall impedance
4	[81]	2-$RC^{\alpha,\beta}$ model	SEI resistance
5	[42]	Warburg model	Fractional order
6	[12]	FO Thevenin model	Resistance
7	[66]	EIS	Fractional order

1. The investigated indexes in Table 2.2 are SOH (No.1 & No.2), RUL (No.3), degradation (No.4 & No.5), and aging level (No.6 & No.7), respectively.

2. Incremental capacity analysis (ICA) in [71] was designed to recognize aging mechanism and estimate SOH in real time.

3. Overall impedance in [20] means the diameter of the semicircular part (in the mid-frequency) of EIS, which would increase due to the increase of charge transfer resistance.

4. SEI (solid electrolyte interface) resistance was observed in [81] and it shows a linear relationship with the remaining capacity, which can indicate the degradation behavior.

5. Fractional order was considered as an indicator of aging level and SOH in [42], which is the research from the same group that has illustrated fractional order can also indicate SOC linearly in [43].

6. In [12], Sutter et al. provided both ohmic resistance and charge transfer resistance evolution over lifetime.

7. In [66], the fractional order α_ω was considered as an indicator of aging level just like that in [42], but the difference is that the variable fractional order α_ω was investigated with the on-line capacitive resistance arc of EIS in frequency domain, which is similar to that in [20]. Hence, the fractional order α_ω was verified to be an index of states and performance of LIBs, like SOH, RUL, etc.

8. Although the SOH estimation, RUL estimation, degradation, and aging problems of LIBs are related with each other, all the No.1–No.6 methods in Table 2.2

were designed specifically for the certain type of index. Hence, all the six kinds of methods would be extended for the other indexes, or for the combining estimation of these indexes. As to the seventh method in [66], it only indicates that fractional order α_ω is relevant to SOC, aging level, and discharge rate of LIBs, but does not provide specific estimation method for aging level. Thus the findings in [66] remain open to more practical applications.

2.5 CHAPTER SUMMARY

This chapter presents a state-of-the-art survey on fractional-order modeling and estimation methods for LIBs mainly in last five years. FO electrochemical models and ECMs are the main model forms for LIBs, and a very detailed presentation and analysis of six types of ECMs in Figure 2.7 are provided in Section 2.3.2. Moreover, all the FO estimation methods for SOC, SOH, and RUL of LIBs are introduced and analyzed in Section 2.4. FO observers and FO-KFs are the two main methods applied to SOC estimation, and eight kinds of FO-KFs are listed in Table 2.1 with brief comments and analysis. While SOH and RUL estimation methods listed in Table 2.2 are not as many as those for SOC estimation, there is still a lot to be investigated for fractional calculus in LIBs lifetime research. The following are some suggestions that may be helpful in future work.

1. Online and real-time monitoring and estimation are the new trends for future BMS; thus, FO modeling, parameter identification, and estimation methods that can work online will have significance for LIBs.

2. Adaptive methods are required for further research, including adaptive parameters of the FO identification algorithm, the weights update of the FO iterative process during estimation, and the updates over the LIBs ageing.

3. Other technologies may be combined with current models and estimation methods. FO filters except FO-KFs, FONN, and FO-EL may be applied for parameter identification and SOC estimation.

4. SOH, RUL, aging, and degradation problems are lack of investigation with fractional calculus, while these problems with integer-order calculus already have many related results that can be referred from [77].

Bibliography

[1] Tareq Abuaisha and Jana Kertzscher. Fractional-order modelling and parameter identification of electrical coils. *Fractional Calculus and Applied Analysis*, 22(1):193–216, 2019.

[2] Avishek Adhikary, Pritin Sen, Siddharha Sen, and Karabi Biswas. Design and performance study of dynamic fractors in any of the four quadrants. *Circuits, Systems, and Signal Processing*, 35(6):1909–1932, 2016.

[3] SMM Alavi, CR Birkl, and DA Howey. Time-domain fitting of battery electro-chemical impedance models. *Journal of Power Sources*, 288:345–352, 2015.

[4] Anis Allagui, Todd J Freeborn, Ahmed S Elwakil, Mohammed E Fouda, Brent J Maundy, Ahmad G Radwan, Zafar Said, and Mohammad Ali Abdelkareem. Review of fractional-order electrical characterization of supercapacitors. *Journal of Power Sources*, 400:457–467, 2018.

[5] Ahmed Alsaedi, Bashir Ahmad, and Mokhtar Kirane. A survey of useful inequalities in fractional calculus. *Fractional Calculus and Applied Analysis*, 20(3):574–594, 2017.

[6] Christoph R Birkl, Matthew R Roberts, Euan McTurk, Peter G Bruce, and David A Howey. Degradation diagnostics for lithium ion cells. *Journal of Power Sources*, 341:373–386, 2017.

[7] Ming Cai, Weijie Chen, and Xiaojun Tan. Battery state-of-charge estimation based on a dual unscented Kalman filter and fractional variable-order model. *Energies*, 10(10):1577, 2017.

[8] Hicham Chaoui, Asmae El Mejdoubi, and Hamid Gualous. Online parameter identification of lithium-ion batteries with surface temperature variations. *IEEE Transactions on Vehicular Technology*, 66(3):2000–2009, 2016.

[9] Hicham Chaoui and Hamid Gualous. Adaptive state of charge estimation of lithium-ion batteries with parameter and thermal uncertainties. *IEEE Transactions on Control Systems Technology*, 25(2):752–759, 2016.

[10] Yixing Chen, Deqing Huang, Qiao Zhu, Weiqun Liu, Congzhi Liu, and Neng Xiong. A new state of charge estimation algorithm for lithium-ion batteries based on the fractional unscented Kalman filter. *Energies*, 10(9):1313, 2017.

[11] Mikaél Cugnet, Jocelyn Sabatier, Stéphane Laruelle, Sylvie Grugeon, Bernard Sahut, Alain Oustaloup, and Jean-Marie Tarascon. On lead-acid-battery resistance and cranking-capability estimation. *IEEE Transactions on Industrial Electronics*, 57(3):909–917, 2009.

[12] Lysander De Sutter, Yousef Firouz, Joris De Hoog, Noshin Omar, and Joeri Van Mierlo. Battery aging assessment and parametric study of lithium-ion batteries by means of a fractional differential model. *Electrochimica Acta*, 305:24–36, 2019.

[13] Achraf Nasser Eddine, Benoît Huard, Jean-Denis Gabano, and Thierry Poinot. Initialization of a fractional order identification algorithm applied for lithium-ion battery modeling in time domain. *Communications in Nonlinear Science and Numerical Simulation*, 59:375–386, 2018.

[14] Bazhlekova Emilia. Subordination in a class of generalized time-fractional diffusion-wave equations. *Fractional Calculus and Applied Analysis*, 21(4):869–900, 2018.

[15] Christian Fleischer, Wladislaw Waag, Hans-Martin Heyn, and Dirk Uwe Sauer. On-line adaptive battery impedance parameter and state estimation considering physical principles in reduced order equivalent circuit battery models part 2. parameter and state estimation. *Journal of Power Sources*, 262:457–482, 2014.

[16] JM Francisco, Jocelyn Sabatier, Loïc Lavigne, F Guillemard, M Moze, M Tari, M Merveillaut, and A Noury. Lithium-ion battery state of charge estimation using a fractional battery model. In *ICFDA'14 International Conference on Fractional Differentiation and Its Applications 2014*, pages 1–6. IEEE, 2014.

[17] Thomas F Fuller, Marc Doyle, and John Newman. Simulation and optimization of the dual lithium ion insertion cell. *Journal of the Electrochemical Society*, 141(1):1–10, 1994.

[18] Jean-Denis Gabano, Thierry Poinot, and Benoît Huard. Bounded diffusion impedance characterization of battery electrodes using fractional modeling. *Communications in Nonlinear Science and Numerical Simulation*, 47:164–177, 2017.

[19] Taedong Goh, Minjun Park, Gyogwon Koo, Minhwan Seo, and Sang Woo Kim. State-of-health estimation algorithm of Li-ion battery using impedance at low sampling rate. In *IEEE PES APPEEC*, pages 146–150, 2016.

[20] Arijit Guha and Amit Patra. Online estimation of the electrochemical impedance spectrum and remaining useful life of lithium-ion batteries. *IEEE Transactions on Instrumentation and Measurement*, 67(8):1836–1849, 2018.

[21] JI Hidalgo-Reyes, José Francisco Gómez-Aguilar, Ricardo Fabricio Escobar-Jiménez, Victor Manuel Alvarado-Martínez, and MG López-López. Classical and fractional-order modeling of equivalent electrical circuits for supercapacitors and batteries, energy management strategies for hybrid systems and methods for the state of charge estimation: A state of the art review. *Microelectronics Journal*, 85:109–128, 2019.

[22] Minghui Hu, Yunxiao Li, Shuxian Li, Chunyun Fu, Datong Qin, and Zonghua Li. Lithium-ion battery modeling and parameter identification based on fractional theory. *Energy*, 165:153–163, 2018.

[23] Xiaosong Hu, Hao Yuan, Changfu Zou, Zhe Li, and Lei Zhang. Co-estimation of state of charge and state of health for lithium-ion batteries based on fractional-order calculus. *IEEE Transactions on Vehicular Technology*, 67(11):10319–10329, 2018.

[24] Ruituo Huai, Zhihao Yu, and Hongyu Li. Historical data demand in window-based battery parameter identification algorithm. *Journal of Power Sources*, 433:126080, 2010.

[25] Pierre E Jacob, Seyed Mohammad Mahdi Alavi, Adam Mahdi, Stephen J Payne, and David A Howey. Bayesian inference in non-Markovian state-space

models with applications to battery fractional-order systems. *IEEE Transactions on Control Systems Technology*, 26(2):497–506, 2017.

[26] Yunfeng Jiang, Bing Xia, Xin Zhao, Truong Nguyen, Chris Mi, and Raymond A de Callafon. Data-based fractional differential models for non-linear dynamic modeling of a lithium-ion battery. *Energy*, 135:171–181, 2017.

[27] Yunfeng Jiang, Bing Xia, Xin Zhao, Truong Nguyen, Chris Mi, and Raymond A de Callafon. Identification of fractional differential models for lithium-ion polymer battery dynamics. *IFAC-PapersOnLine*, 50(1):405–410, 2017.

[28] Dae-Keun Kang and Heon-Cheol Shin. Investigation on cell impedance for high-power lithium-ion batteries. *Journal of Solid State Electrochemistry*, 11(10):1405–1410, 2007.

[29] Changpin Li and Min Cai. *Theory and Numerical Approximations of Fractional Integrals and Derivatives*. SIAM, 2019.

[30] Changpin Li and Qian Yi. Modeling and computing of fractional convection equation. *Communications on Applied Mathematics and Computation*, pages 1–31, 2019.

[31] Changpin Li, Qian Yi, and Júrgen Kurths. Fractional convection. *Journal of Computational and Nonlinear Dynamics*, 13(1):011004, 2018.

[32] Changpin Li and Fanhai Zeng. *Numerical Methods for Fractional Calculus*. Chapman and Hall/CRC, 2015.

[33] Shuxian Li, Minghui Hu, Yunxiao Li, and Changchao Gong. Fractional-order modeling and SOC estimation of lithium-ion battery considering capacity loss. *International Journal of Energy Research*, 43(1):417–429, 2019.

[34] Xiaoyu Li, Guodong Fan, Ke Pan, Guo Wei, Chunbo Zhu, Giorgio Rizzoni, and Marcello Canova. A physics-based fractional order model and state of energy estimation for lithium ion batteries. Part I: Model development and observability analysis. *Journal of Power Sources*, 367:187–201, 2017.

[35] Xiaoyu Li, Ke Pan, Guodong Fan, Rengui Lu, Chunbo Zhu, Giorgio Rizzoni, and Marcello Canova. A physics-based fractional order model and state of energy estimation for lithium ion batteries. Part II: Parameter identification and state of energy estimation for LiFePO4 battery. *Journal of Power Sources*, 367:202–213, 2017.

[36] Yanwen Li, Chao Wang, and Jinfeng Gong. A wavelet transform-adaptive unscented Kalman filter approach for state of charge estimation of LiFePo4 battery. *International Journal of Energy Research*, 42(2):587–600, 2018.

[37] Congzhi Liu, Weiqun Liu, Lingyan Wang, Guangdi Hu, Luping Ma, and Bingyu Ren. A new method of modeling and state of charge estimation of the battery. *Journal of Power Sources*, 320:1–12, 2016.

[38] Datong Liu, Wei Xie, Haitao Liao, and Yu Peng. An integrated probabilistic approach to lithium-ion battery remaining useful life estimation. *IEEE Transactions on Instrumentation and Measurement*, 64(3):660–670, 2014.

[39] Datong Liu, Jianbao Zhou, Dawei Pan, Yu Peng, and Xiyuan Peng. Lithium-ion battery remaining useful life estimation with an optimized Relevance Vector Machine algorithm with incremental learning. *Measurement*, 63:143–151, 2015.

[40] Kailong Liu, Kang Li, Qiao Peng, and Cheng Zhang. A brief review on key technologies in the battery management system of electric vehicles. *Frontiers of Mechanical Engineering*, 14(1):47–64, 2019.

[41] Shulin Liu, Xia Dong, and Yun Zhang. A new state of charge estimation method for lithium-ion battery based on the fractional order model. *IEEE Access*, 7:122949–122954, 2019.

[42] Xin Lu, Hui Li, and Ning Chen. An indicator for the electrode aging of lithium-ion batteries using a fractional variable order model. *Electrochimica Acta*, 299:378–387, 2019.

[43] Xin Lu, Hui Li, Jun Xu, Siyuan Chen, and Ning Chen. Rapid estimation method for state of charge of lithium-ion battery based on fractional continual variable order model. *Energies*, 11(4):714, 2018.

[44] Jiayi Luo, Jiankun Peng, and Hongwen He. Lithium-ion battery SOC estimation study based on Cubature Kalman filter. *Energy Procedia*, 158:3421–3426, 2019.

[45] Yan Ma, Xiuwen Zhou, Bingsi Li, and Hong Chen. Fractional modeling and SOC estimation of lithium-ion battery. *IEEE/CAA Journal of Automatica Sinica*, 3(3):281–287, 2016.

[46] JA Tenreiro Machado and Virginia Kiryakova. The chronicles of fractional calculus. *Fractional Calculus and Applied Analysis*, 20(2):307–336, 2017.

[47] JA Tenreiro Machado and António M Lopes. Fractional state space analysis of temperature time series. *Fractional Calculus and Applied Analysis*, 18(6):1518–1536, 2015.

[48] Egoitz Martinez-Laserna, Elixabet Sarasketa-Zabala, Igor Villarreal Sarria, Daniel-Ioan Stroe, Maciej Swierczynski, Alexander Warnecke, Jean-Marc Timmermans, Shovon Goutam, Noshin Omar, and Pedro Rodriguez. Technical viability of battery second life: a study from the ageing perspective. *IEEE Transactions on Industry Applications*, 54(3):2703–2713, 2018.

[49] Kodjo SR Mawonou, Akram Eddahech, Didier Dumur, Dominique Beauvois, and Emmanuel Godoy. Improved state of charge estimation for li-ion batteries using fractional order extended Kalman filter. *Journal of Power Sources*, 435:226710, 2019.

[50] Sara Mohajer, Jocelyn Sabatier, Patrick Lanusse, and Olivier Cois. A fractional-order electro-thermal aging model for lifetime enhancement of lithium-ion batteries. *IFAC-PapersOnLine*, 51(2):220–225, 2018.

[51] Maxime Montaru and Serge Pelissier. Frequency and temporal identification of a Li-ion polymer battery model using fractional impedance. *Oil & Gas Science and Technology–Revue de l'Institut Français du Pétrole*, 65(1):67–78, 2010.

[52] Hao Mu, Rui Xiong, Hongfei Zheng, Yuhua Chang, and Zeyu Chen. A novel fractional order model based state-of-charge estimation method for lithium-ion battery. *Applied Energy*, 207:384–393, 2017.

[53] Arkadiusz Mystkowski and Argyrios Zolotas. PLC-based discrete fractional-order control design for an industrial-oriented water tank volume system with input delay. *Fractional Calculus and Applied Analysis*, 21(4):1005–1026, 2018.

[54] Achraf Nasser-Eddine, Benoît Huard, Jean-Denis Gabano, and Thierry Poinot. A two steps method for electrochemical impedance modeling using fractional order system in time and frequency domains. *Control Engineering Practice*, 86:96–104, 2019.

[55] HE Perez, S Dey, X Hu, and SJ Moura. Optimal charging of Li-ion batteries via a single particle model with electrolyte and thermal dynamics. *Journal of The Electrochemical Society*, 164(7):A1679–A1687, 2017.

[56] Humberto Rafeiro and Stefan Samko. Fractional integrals and derivatives: mapping properties. *Fractional Calculus and Applied Analysis*, 19(3):580–607, 2016.

[57] Saeed Khaleghi Rahimian, Sean Rayman, and Ralph E White. Extension of physics-based single particle model for higher charge–discharge rates. *Journal of Power Sources*, 224:180–194, 2013.

[58] J. Rifkin. The third industrial revolution: a radical new sharing economy. *https://www.singularityweblog.com/third-industrial-revolution/*, 2018.

[59] Jocelyn Sabatier, Mohamed Aoun, Alain Oustaloup, Gilles Grégoire, Franck Ragot, and Patrick Roy. Fractional system identification for lead acid battery state of charge estimation. *Signal Processing*, 86(10):2645–2657, 2006.

[60] Jocelyn Sabatier, Mikael Cugnet, Stephane Laruelle, Sylvie Grugeon, Bernard Sahut, A Oustaloup, and JM Tarascon. A fractional order model for lead-acid battery crankability estimation. *Communications in Nonlinear Science and Numerical Simulation*, 15(5):1308–1317, 2010.

[61] Jocelyn Sabatier, Junior Mbala Francisco, Franck Guillemard, Loic Lavigne, Mathieu Moze, and Mathieu Merveillaut. Lithium-ion batteries modeling: A simple fractional differentiation based model and its associated parameters estimation method. *Signal Processing*, 107:290–301, 2015.

[62] Jocelyn Sabatier, Franck Guillemard, Loic Lavigne, Agnieszka Noury, Mathieu Merveillaut, and Junior Mbala Francico. Fractional models of lithium-ion batteries with application to state of charge and ageing estimation. In *Informatics in Control, Automation and Robotics*, pages 55–72. Springer, 2018.

[63] Jocelyn Sabatier, Mathieu Merveillaut, Junior Mbala Francisco, Franck Guillemard, and Denis Porcelatto. Fractional models for lithium-ion batteries. In *European Control Conference*, pages 3458–3463, 2013.

[64] Jocelyn Sabatier, Mathieu Merveillaut, Junior Mbala Francisco, Franck Guillemard, and Denis Porcelatto. Lithium-ion batteries modeling involving fractional differentiation. *Journal of Power Sources*, 262:36–43, 2014.

[65] HongGuang Sun, Ailian Chang, Yong Zhang, and Wen Chen. A review on variable-order fractional differential equations: mathematical foundations, physical models, numerical methods and applications. *Fractional Calculus and Applied Analysis*, 22(1):27–59, 2019.

[66] Yue Sun, Yan Li, Meijuan Yu, Zhongkai Zhou, Qi Zhang, Bin Duan, Yunlong Shang, and Chenghui Zhang. Variable fractional order-a comprehensive evaluation indicator of lithium-ion batteries. *Journal of Power Sources*, page 227411, 2019.

[67] P. T. Systems. Lithium-ion battery advantages. *https://www.powertechsystems. eu/home/tech-corner/lithium-ion-battery-advantages/*, 2019.

[68] Takahiro Takamatsu and Hiromitsu Ohmori. State and parameter estimation of lithium-ion battery by Kreisselmeier-type adaptive observer for fractional calculus system. In *54th Annual Conference SICE of Japan*, pages 86–90, 2015.

[69] Takahiro Takamatsu and Hiromitsu Ohmori. Online parameter estimation for lithium-ion battery by using adaptive observer for fractional-order system. *Electronics and Communications in Japan*, 101(3):80–89, 2018.

[70] Xiaopeng Tang, Yujie Wang, Changfu Zou, Ke Yao, Yongxiao Xia, and Furong Gao. A novel framework for lithium-ion battery modeling considering uncertainties of temperature and aging. *Energy Conversion and Management*, 180:162–170, 2019.

[71] Jinpeng Tian, Rui Xiong, and Quanqing Yu. Fractional-order model-based incremental capacity analysis for degradation state recognition of lithium-ion batteries. *IEEE Transactions on Industrial Electronics*, 66(2):1576–1584, 2018.

[72] Baojin Wang, Shengbo Eben Li, Huei Peng, and Zhiyuan Liu. Fractional-order modeling and parameter identification for lithium-ion batteries. *Journal of Power Sources*, 293:151–161, 2015.

[73] Baojin Wang, Zhiyuan Liu, Shengbo Eben Li, Scott Jason Moura, and Huei Peng. State-of-charge estimation for lithium-ion batteries based on a nonlinear

fractional model. *IEEE Transactions on Control Systems Technology*, 25(1):3–11, 2016.

[74] Chuanxin Wang, Qin Huang, and Rui Ling. Battery SOC estimating using a fractional order unscented Kalman filter. In *Chinese Automation Congress*, pages 1268–1273, 2015.

[75] Jianlin Wang, Le Zhang, Dan Xu, Peng Zhang, and Gairu Zhang. A simplified fractional order equivalent circuit model and adaptive online parameter identification method for lithium-ion batteries. *Mathematical Problems in Engineering*, 2019, 2019.

[76] Yebin Wang, Huazhen Fang, Lei Zhou, and Toshihiro Wada. Revisiting the state-of-charge estimation for lithium-ion batteries: A methodical investigation of the extended Kalman filter approach. *IEEE Control Systems Magazine*, 37(4):73–96, 2017.

[77] Nikolaos Wassiliadis, Jorn Adermann, Alexander Frericks, Mikhail Pak, Christoph Reiter, Boris Lohmann, and Markus Lienkamp. Revisiting the dual extended Kalman filter for battery state-of-charge and state-of-health estimation: A use-case life cycle analysis. *Journal of Energy Storage*, 19:73–87, 2018.

[78] Zhongbao Wei, Binyu Xiong, Dongxu Ji, and King Jet Tseng. Online state of charge and capacity dual estimation with a multi-timescale estimator for lithium-ion battery. *Energy Procedia*, 105:2953–2958, 2017.

[79] Renxin Xiao, Jiangwei Shen, Xiaoyu Li, Wensheng Yan, Erdong Pan, and Zheng Chen. Comparisons of modeling and state of charge estimation for lithium-ion battery based on fractional order and integral order methods. *Energies*, 9(3):184, 2016.

[80] Rui Xiong, Jiayi Cao, Quanqing Yu, Hongwen He, and Fengchun Sun. Critical review on the battery state of charge estimation methods for electric vehicles. *IEEE Access*, 6:1832–1843, 2017.

[81] Rui Xiong, Jinpeng Tian, Hao Mu, and Chun Wang. A systematic model-based degradation behavior recognition and health monitoring method for lithium-ion batteries. *Applied Energy*, 207:372–383, 2017.

[82] Rui Xiong, Jinpeng Tian, Weixiang Shen, and Fengchun Sun. A novel fractional order model for state of charge estimation in lithium ion batteries. *IEEE Transactions on Vehicular Technology*, 68(5):4130–4139, 2018.

[83] Rami Yamin and Ahmed Rachid. Embedded state of charge and state of health estimator based on Kalman filter for electric scooter battery management system. In *IEEE ICCE-Berlin*, pages 440–444, 2014.

[84] Qingxia Yang, Jun Xu, Binggang Cao, and Xiuqing Li. A simplified fractional order impedance model and parameter identification method for lithium-ion batteries. *PloS one*, 12(2):e0172424, 2017.

[85] Min Ye, Hui Guo, and Binggang Cao. A model-based adaptive state of charge estimator for a lithium-ion battery using an improved adaptive particle filter. *Applied Energy*, 190:740–748, 2017.

[86] Shifei Yuan, Hongjie Wu, Xi Zhang, and Chengliang Yin. Online estimation of electrochemical impedance spectra for lithium-ion batteries via discrete fractional order model. In *2013 IEEE Vehicle Power and Propulsion Conference (VPPC)*, pages 1–6. IEEE, 2013.

[87] Chenghui Zhang, Yun Zhang, and Yan Li. A novel battery state-of-health estimation method for hybrid electric vehicles. *IEEE/ASME Transactions On Mechatronics*, 20(5):2604–2612, 2015.

[88] Lei Zhang, Xiaosong Hu, Zhenpo Wang, Fengchun Sun, and David G Dorrell. A review of supercapacitor modeling, estimation, and applications: A control/management perspective. *Renewable and Sustainable Energy Reviews*, 81:1868–1878, 2018.

[89] Qi Zhang, Naxin Cui, Yunlong Shang, Guojing Xing, and Chenghui Zhang. Relevance between fractional-order hybrid model and unified equivalent circuit model of electric vehicle power battery. *Science China Information Sciences*, 61(7):70208–1, 2018.

[90] Qi Zhang, Yan Li, Yunlong Shang, Bin Duan, Naxin Cui, and Chenghui Zhang. A fractional-order kinetic battery model of lithium-ion batteries considering a nonlinear capacity. *Electronics*, 8(4):394, 2019.

[91] Qi Zhang, Yunlong Shang, Yan Li, Naxin Cui, Bin Duan, and Chenghui Zhang. A novel fractional variable-order equivalent circuit model and parameter identification of electric vehicle Li-ion batteries. *ISA Transactions*, 97:448–457, 2020.

[92] Ruifeng Zhang, Bizhong Xia, Baohua Li, Libo Cao, Yongzhi Lai, Weiwei Zheng, Huawen Wang, and Wei Wang. State of the art of lithium-ion battery SOC estimation for electrical vehicles. *Energies*, 11(7):1820, 2018.

[93] Yang Zhao, Yan Li, Fengyu Zhou, Zhongkai Zhou, and YangQuan Chen. An iterative learning approach to identify fractional order KiBaM model. *IEEE/CAA Journal of Automatica Sinica*, 4(2):322–331, 2017.

[94] Fangdan Zheng, Yinjiao Xing, Jiuchun Jiang, Bingxiang Sun, Jonghoon Kim, and Michael Pecht. Influence of different open circuit voltage tests on state of charge online estimation for lithium-ion batteries. *Applied Energy*, 183:513–525, 2016.

[95] Fuli Zhong, Hui Li, and Qishui Zhong. An approach for SOC estimation based on sliding mode observer and fractional order equivalent circuit model of lithium-ion batteries. In *2014 IEEE International Conference on Mechatronics and Automation*, pages 1497–1503. IEEE, 2014.

[96] Fuli Zhong, Hui Li, Shouming Zhong, Qishui Zhong, and Chun Yin. An SOC estimation approach based on adaptive sliding mode observer and fractional order equivalent circuit model for lithium-ion batteries. *Communications in Nonlinear Science and Numerical Simulation*, 24(1-3):127–144, 2015.

[97] Qishui Zhong, Fuli Zhong, Jun Cheng, Hui Li, and Shouming Zhong. State of charge estimation of lithium-ion batteries using fractional order sliding mode observer. *ISA Transactions*, 66:448–459, 2017.

[98] Daming Zhou, Ke Zhang, Alexandre Ravey, Fei Gao, and Abdellatif Miraoui. Parameter sensitivity analysis for fractional-order modeling of lithium-ion batteries. *Energies*, 9(3):123, 2016.

[99] Qiao Zhu et al. A state of charge estimation approach based on fractional order adaptive extended Kalman filter for lithium-ion batteries. In *IEEE 7th DDCLS*, pages 271–276, 2018.

[100] Changfu Zou, Xiaosong Hu, Satadru Dey, Lei Zhang, and Xiaolin Tang. Nonlinear fractional-order estimator with guaranteed robustness and stability for lithium-ion batteries. *IEEE Transactions on Industrial Electronics*, 65(7):5951–5961, 2017.

[101] Changfu Zou, Lei Zhang, Xiaosong Hu, Zhenpo Wang, Torsten Wik, and Michael Pecht. A review of fractional-order techniques applied to lithium-ion batteries, lead-acid batteries, and supercapacitors. *Journal of Power Sources*, 390:286–296, 2018.

[102] Yuan Zou, Shengbo Eben Li, Bing Shao, and Baojin Wang. State-space model with non-integer order derivatives for lithium-ion battery. *Applied Energy*, 161:330–336, 2016.

CHAPTER 3

Fractional-Order Algorithms for Battery State Estimation

To address the problem of the inability to explain the black box machine learning algorithms and to achieve the integration of battery mechanisms and intelligent estimation algorithms, this chapter explores the transformation of battery electrochemical characteristics into physical mathematical constraints that can be used to guide the training and convergence process of machine learning algorithms, attempting to use battery mechanisms to assist intelligent algorithms in leveraging battery physics knowledge. This chapter takes neural networks as the basic algorithm architecture, and combines the optimization method and loss function to form the network embedded with battery physics knowledge. Using the algorithm idea of physical information constrained machine learning, the fractional gradient descent method and battery fractional-order loss function are proposed. A physical information constrained neural network that integrates battery voltage variation law is constructed for lithium-ion battery state estimation. The physical information constrained neural network algorithm proposed in this chapter provides ideas for the transformation of machine learning from black box models to gray box models, and provides an effective and feasible fusion of battery mechanism and machine learning. At the same time, it also improves the accuracy of lithium-ion battery state estimation and provides a new algorithm for state estimation.

3.1 PHYSICS-INFORMED CONDITIONS FOR BATTERY

3.1.1 Physics-informed machine learning

At present, model-based methods for describing battery mechanisms (white box) and AI-based data-driven learning methods for batteries (black box) are developed in two relatively independent directions in research. The existing machine learning algorithms are designed from the perspectives of statistics and data probability theory. For the electrochemical internal states of lithium-ion batteries, such as loss of lithium-ion (LLI) and loss of active material (LAM) [5], they cannot be mapped using existing advanced machine learning algorithms. Therefore, it is necessary to enhance the interpretability of the current algorithm to deeply learn the internal information

DOI: 10.1201/9781003670902-3

of lithium-ion batteries, and physics-informed machine learning (PIML) provides a feasible paradigm for this. The concept of physical information constrained machine learning was proposed in as early as 2017 [8], and the related research has exploded since then. A physics-based machine learning approach was proposed to accelerate training and enhance machine learning by embedding physics, such as observation bias, induction bias, and learning bias [8], as shown in Figure 3.1.

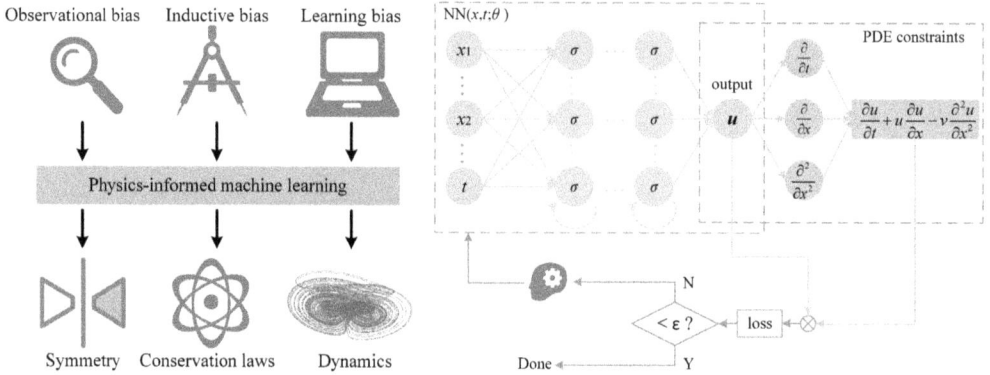

Figure 3.1 Physics-informed machine learning (PIML), adapted/paraphased from [8].

Part of the parameters of the pseudo-two-dimensions (P2D) mechanism model and the tuning of neural network parameters do not often have unique solutions. In neural networks, there are also optimization-related issues such as features, structures, and parameters (Figure 3.2). It is particularly important to determine the relationship and correlation strength between these issues in order to obtain an effective optimization direction.

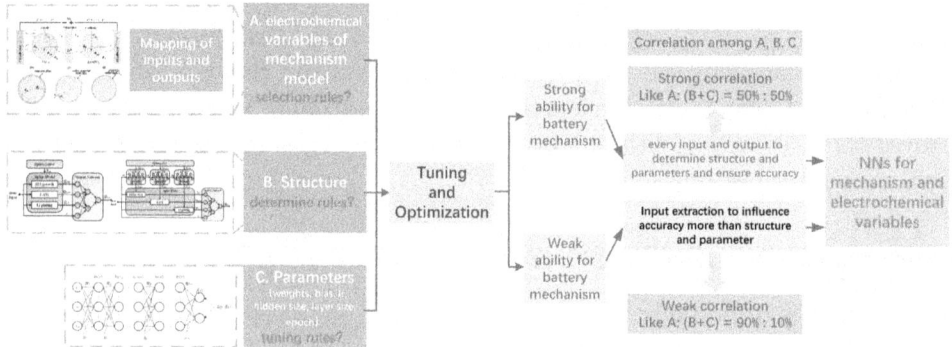

Figure 3.2 Optimization of features and structures for mechanism interpretability.

When designing physical information constraint modeling algorithms for lithium-ion batteries, due to the fractional order characteristics of lithium-ion batteries, some researchers have studied fractional-order recurrent neural network (FORNN) theory to process battery time series data [6, 7]. However, most of the current literature focuses on theoretical derivation and stability analysis. In addition, some researchers have proposed physics-based deep learning [19] or physics-based neural networks [11] for state estimation of lithium-ion batteries, but current research mainly focuses on optimizing machine learning feature inputs using battery mechanisms.

In this chapter, we will combine the fractional-order characteristics of lithium-ion batteries with the interpretability design of neural network algorithms, and propose a physics-informed neural network (PINN) with fractional order constraints for high-precision state estimation of lithium-ion batteries. The fractional derivative equation representing the fractional-order law of lithium-ion batteries is introduced into the fractional-order gradient descent (FOGD) method and loss function of PINN, to guide the backpropagation computation of gradient and weight updates during the training process. The battery information is encoded into the algorithm to achieve feature learning of the neural network. The battery mechanism is fused with the neural network algorithm to construct a machine learning framework (gray box) for knowledge embedding to enhance the interpretability of the algorithm. This chapter conducts SOC estimation experiments under Federal Urban Driving Conditions (FUDS) and SOH estimation experiments under real vehicle operation big data to verify the performance of the proposed PINN. It is worth mentioning that only SOC estimation and SOH estimation were conducted in this chapter, but the proposed algorithm paradigm can be extended to other state estimation of lithium-ion batteries.

3.1.2 Fractional-order characteristics of lithium-ion battery

3.1.2.1 Fractional-order gradient

Three commonly used definitions for fractional derivatives are Riemann-Liouville (R-L), Grünwald-Letnikov (G-L), and Caputo [1]. R-L definition has too complicated initial conditions to calculate in applications; thus, this chapter only considers G-L definition and Caputo definition.

Grünwald-Letnikov (G-L) derivative [21].

$$^{GL}D_t^\alpha f(t) = \lim_{h \to 0} h^{-\alpha} \sum_{j=0}^{[\frac{t-t_0}{h}]} (-1)^j \binom{\alpha}{j} f(t - jh), \tag{3.1}$$

where $^{GL}D_t^\alpha$ represents the fractional operator in G-L sense, $f(t)$ is a certain integrable function in $[a, b]$, $[\frac{t-t_0}{h}]$ represents the number of the approximate recursive terms of integer order parts, and $\binom{\alpha}{j} = \alpha!/j!(\alpha - j)!$ represents the coefficients of the approximate recursive terms.

Finite form of Grünwald-Letnikov derivative.
If the infinite form in (3.1) is limited as finite terms L, $^{GL}D_t^\alpha$ can be approximated as [14]

$$\Delta^\alpha f(x) = \sum_{j=0}^{L} (-1)^j \binom{\alpha}{j} f(k - j), \tag{3.2}$$

where the sum formula in the right side only selects L finite terms of $f(k - j), j = 0, 1, ..., L$, which are the discrete values in the L previous sampling moments.

Caputo derivative [21] is defined as

$$^{C}D_t^{\alpha} f(t) = \frac{1}{\Gamma(n-\alpha)} \int_{t_0}^{t} \frac{f^{(n)}(\tau)}{(t-\tau)^{\alpha-n+1}} d\tau, t > t_0, \tag{3.3}$$

where $^{C}D_x^{\alpha}$ represents the fractional operator in Caputo sense, $\alpha \in (n-1, n)$, $\Gamma(\cdot)$ is the Gamma function.

If $f(x)$ is sufficiently smooth, Caputo derivative in (3.3) can be discretized as [13]

$$^{C}_{x_0}D_x^{\alpha} f(x) = \sum_{i=1}^{\infty} \frac{f^{(i)}(x_0)}{\Gamma(i+1-\alpha)}(x-x_0)^{i-\alpha}. \tag{3.4}$$

In (3.4), if $f(x) = (x-x_0)^{\lambda}$, (3.4) can be deduced as [9]

$$^{C}_{x_0}D_x^{\alpha}(x-x_0)^{\lambda} = \frac{\Gamma(\lambda+1)}{\Gamma(\lambda-\alpha+1)}(x-x_0)^{\lambda-\alpha}, \tag{3.5}$$

where $\lambda - \alpha + 1 > 0$.

G-L definition in (3.2) has a discrete implement form for application, and Caputo definition in (3.5) holds a simple form for power function. As more general forms of derivative than the integer-order one, both G-L definition and Caputo definition can be transformed into the integer-order derivative when $\alpha = 1$. Hence, we employ G-L derivative for fractional-order state feedback and fractional-order constraints, and Caputo derivative for backpropagation with fractional-order gradient in this chapter.

Consider an unconstrained convex optimization problem [2]

$$\min_{x} f(x), \tag{3.6}$$

where $f(x)$ is a smooth convex function with a unique global extremum point x^*. Gradient in continuous form can be presented as

$$\dot{x} = -\rho \nabla f(x), \tag{3.7}$$

where $\rho > 0$ is the step size, and $\nabla f(x)$ is the gradient of $f(x)$ at x. Equation (3.7) can be discretized as

$$x_{k+1} = x_k - \rho \nabla f(x_k), \tag{3.8}$$

where k is the iteration step, and x_k is the discrete value at the step k. Fractional-order gradient in Caputo definition can be presented as

$$x_{k+1} = x_k - \rho \nabla_{x_k}^{C} D_{x_{k+1}}^{\alpha} f(x), \tag{3.9}$$

where $0 < \alpha < 1$. The convergence of fractional-order gradient in (3.9) depends on the fractional order α and the initial value x_0, and (3.9) can converge to global extremum point x^* if Caputo definition is calculated by the discrete equation (3.4). In this chapter, fractional-order gradients are embedded into the training process of FORNN to accelerate training process and improve prediction accuracy with battery physics information.

3.1.2.2 Fractional-order equivalent circuit model (ECM)

Compared to electrochemical model, ECM for LIB holds simpler structure but enough battery physics information for application to NNs. Fractional order constant phase element (CPE), or Warburg element, is introduced due to the capacitance of LIBs in low-frequency and mid-frequency ranges are not ideal first-order derivative, and should be modeled as a fractional-order element [21]. Hence, in this chapter, LIB's fractional-order modeling (FOM) is employed to describe the three main parts (high-frequency inductive tail, mid-frequency reaction, and low-frequency diffusion dynamics) of the battery electrochemical impedance spectroscopy (EIS) [16].

Warburg element or CPE can manifest the phenomenon of capacitance dispersion instead of an ideal capacitor in LIBs [21]. The voltage-current relationship in time domain and the impedance in frequency domain of a CPE is presented as

$$\begin{cases} i(t) = C_{CPE}\frac{d^\alpha u(t)}{dt^\alpha}, & 0 < \alpha < 1, t \geq 0 \\ Z_{CPE}(s) = \frac{U(s)}{I(s)} = \frac{1}{C_{CPE} \cdot s^\alpha} = \frac{1}{C_{CPE}(j\omega)^\alpha} \end{cases}, \tag{3.10}$$

where Z_{CPE} is the complex impedance, C_{CPE} is the capacity coefficient, j is the imaginary unit, α is the fractional order related to capacitance dispersion, and ω is the angular frequency.

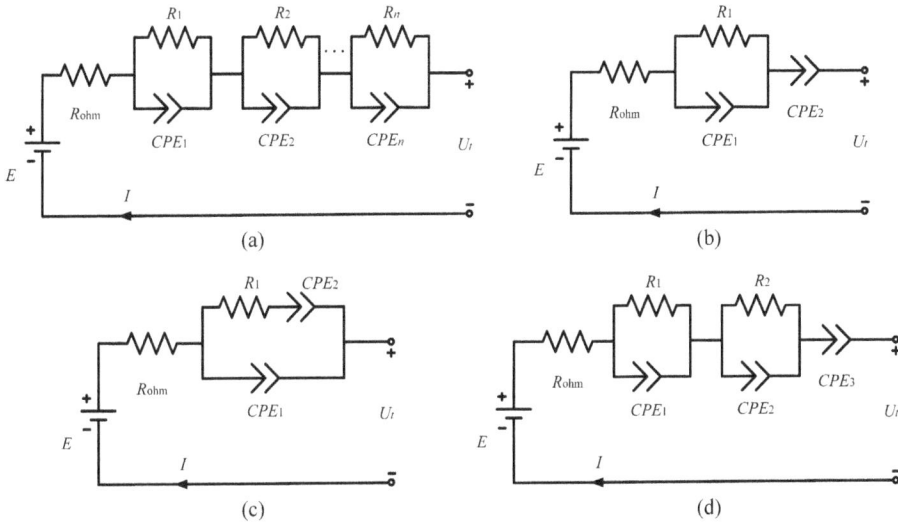

Figure 3.3 Fractional-order equivalent circuit model of LIBs in various forms. (a) $n - RC^\alpha$ Model, (b) fractional-order PNGV model, (c) fractional-order Randles model, (d) three-orders fractional-order model ($CPE_i, i = 1, 2, ..., n$).

Figure 3.3 presents the commonly used fractional-order ECMs in four different forms. If $n = 1$ in Figure 3.3 (a), ECM in Figure 3.3 (a) becomes a fractional-order Thevenin model (also called 1-RC model), which only considers high-frequency and mid-frequency reactions, and ignores Warburg element of diffusion effects [15]. Figure 3.3 (b) and Figure 3.3 (c) are fractional-order Partnership for a New Generation of Vehicles (PNGV) and fractional-order "Randles" model [12, 20], respectively, and both are systems with two fractional orders. Fractional-order Randles model in Figure 3.3 (c) is proposed to mainly reflect the double layer effect in mid-frequency, which shows like a semi-ellipse rather than a semi-circle due to capacitance dispersion, while the fractional-order PNGV model in Figure 3.3 (b) is a widely used FOM due to the full-scale reflection of LIB dynamics in all frequency range. The corresponding pseudo state-space model (SSM) of fractional-order PNGV model is [16]

$$
\begin{bmatrix} \frac{d^{\alpha_1} U_{CPE_1}}{dt^{\alpha_1}} \\ \frac{d^{\alpha_2} U_{CPE_2}}{dt^{\alpha_2}} \end{bmatrix} = \begin{bmatrix} -\frac{1}{R_1 C_{CPE_1}} & 0 \\ 0 & 0 \end{bmatrix} \begin{bmatrix} U_{CPE_1} \\ U_{CPE_2} \end{bmatrix} + \begin{bmatrix} \frac{1}{C_{CPE_1}} \\ \frac{1}{C_{CPE_2}} \end{bmatrix} I,
$$
$$
U_t = \begin{bmatrix} 1 & 1 \end{bmatrix} \begin{bmatrix} U_{CPE_1} \\ U_{CPE_2} \end{bmatrix} + R_{\text{ohm}} I + E,
$$
(3.11)

where U_{CPE_2} and U_{CPE_1} are the voltages of Warburg element CPE_2 and CPE_1; C_{CPE_2} and C_{CPE_1} are the capacitance of CPE_2 and CPE_1; α_1 and α_2 are the fractional orders of CPE_2 and CPE_1, respectively; I is the current. For higher modeling accuracy, some extended high-order FOMs for LIB are proposed by adding low-frequency component and replacing ideal capacitor with CPEs [10], which results in a high-order FOM with three fractional orders as shown in Figure 3.3 (d).

3.2 FRACTIONAL-ORDER RECURRENT NEURAL NETWORK

This section introduces a new interpretation method using the neural network black box model for lithium batteries and constructs a fractional-order prediction "gray box" model to enhance interpretability. Among various neural networks, Recurrent Neural Network (RNN) has a significant predictive capability on objects with temporal features, while lithium battery voltage, current, and temperature are all temporal feature data. To make algorithm informed by physics and battery mechanism, fractional-order derivative is applied to recurrent neural network to construct fractional-order recurrent neural network with physics-informed knowledge , simplified as **FORNN with PIBatKnow** in the following. The integral architecture of FORNN with PIBatKnow is presented in Figure 3.4. Three aspects of FORNN with PIBatKnow can be improved as fractional-order state feedback for forward propagation, fractional-order constraints for loss function, and fractional-order gradients by backpropagation computation, which are introduced in the following subsection, respectively.

The basis of FORNN with PIBatKnow in Figure 3.4 is an RNN, which contains an input layer with m neurons (first layer in Figure 3.4), several hidden layers $(h_1, ..., h_{p-1}, h_p, ..., h_l)$ with $n_p(p = 1, 2, ..., l)$ neurons (the middle part in Figure 3.4), and an output layer with q neurons (last layer in Figure 3.4), respectively.

Suppose that the training input dataset is $(\mathbf{x}_i, \mathbf{y}_i), (i = 1, 2, ..., N)$, where $\mathbf{x}_i = (x_{i1}, x_{i2}, ..., x_{im})^T$ is the network input and $\mathbf{y}_i = (y_{i1}, y_{i2}, ..., y_{iq})^T$ is the network target (training labels). To simplify expression, vectors $\mathbf{x}_i$ and $\mathbf{y}_i$ are presented as

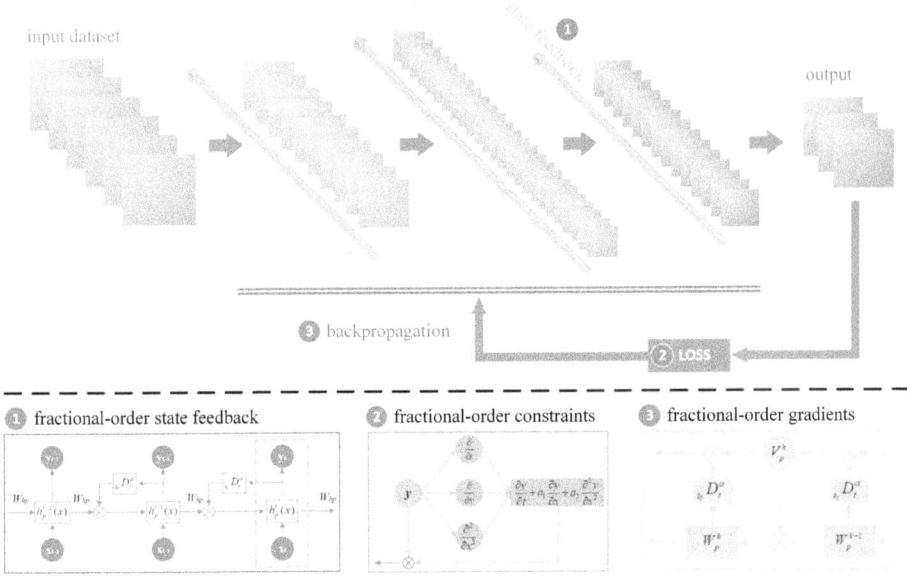

Figure 3.4 Fractional-order recurrent neural network (FORNN) encoding with physics-informed battery knowledge.

x and y in the rest of this section. The specific part of RNN is the chain structure inside the hidden layers, which hold hidden states feedback in time series. Assume W_p and b_p be the weight and bias matrix connecting the $(p-1)$th hidden layer to the pth hidden layer, W_{hp} be the weights for memory updates of the pth hidden layer in the chain structure of RNN, $g(\cdot)$ be the activation functions, and $L(g(x), y)$ be the loss function. Within the pre-set epochs threshold of training process, RNN would go through forward propagation and backpropagation process with training data, and the forward propagation starting from the input layer can be presented as

$$\begin{cases} a_p(x) = W_p h_{p-1}(x) + b_p, \\ h_p(x) = g(a_p(x)), p = 1, 2, ..., l \end{cases} \tag{3.12}$$

where $a_p(x)$ and $h_p(x)$ are the input and the output of the pth hidden layer, respectively. Equation (3.9) is the basic iterative equation of the proposed FORNN with PIBatKnow.

3.2.1 Fractional-order state feedback

Fractional-order derivative can be introduced into the state feedback in the chain structure of the proposed FORNN, resulting in fractional-order state feedback, as shown in Figure 3.5 (b). Figure 3.5 (a) is the integer-order state feedback, which can be presented as

$$_{t_0}D_t h_p(t) = -W_{hp} h_p(t) + W'_p g(h_{p-1}(t)) + d_p, p = 1, 2, ..., l \tag{3.13}$$

where W'_p is a coefficient related with weights W_p, d_p is a constant related with bias b_p, and memory weights W_{hp} are shared among all the moments in time series.

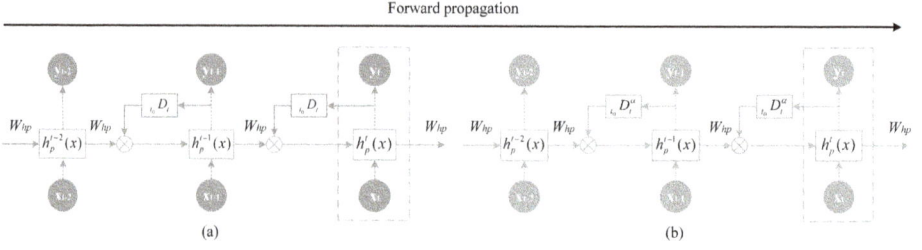

Figure 3.5 Fractional-order state feedback in the forward propagation of recurrent neural network. (a) integer-order state feedback, (b) fractional-order state feedback.

Fractional-order state feedback in Figure 3.5 (b) is extended from the integer-order one in Figure 3.5 (a), thus fractional-order state feedback can be presented as

$$_{t_0}D_t^\alpha h_p(t) = -W_{hp}h_p(t) + W'_p g(h_{p-1}(t)) + d_p, p = 1, 2, ..., l \qquad (3.14)$$

where $_{t_0}D_t^\alpha$ is the fractional-order derivative of $h_p(t)$ in $[t_0, t]$, and (3.14) reduces to the integer-order derivative (3.13) when $\alpha = 1$. Equation (3.14) can be discretized by G-L definition in (3.2) which is

$$\Delta^\alpha h_p(t) = \sum_{j=0}^{L}(-1)^j \begin{pmatrix} \alpha \\ j \end{pmatrix} h_p(k - j) \qquad (3.15)$$

Taking (3.15) into (3.14), we obtain that

$$\sum_{j=0}^{L}(-1)^j \begin{pmatrix} \alpha \\ j \end{pmatrix} h_p(k - j) = -W_{hp}(k)h_p(k - 1) + W'_p(k)g(h_{p-1}(k)) + d_p. \qquad (3.16)$$

Note that $h_p(k)$ is included in the first term ($j = 0$) of (3.15), and the second term of (3.15) is $-h_p(k - 1)$ when $j = 1$. Hence, expanding the first and the second terms in (3.15), then normalizing the coefficients, the discrete form of fractional-order state feedback can be deduced as

$$\begin{aligned} h_p(k) = &(1 - W_{hp}(k))h_p(k - 1) + \sum_{j=2}^{L}(-1)^{j+1} \begin{pmatrix} \alpha \\ j \end{pmatrix} h_p(k - j) \\ &+ W'_p(k)g(h_{p-1}(k)) + d_p \end{aligned} \qquad (3.17)$$

The upper limit L of the finite discrete form in (3.17) should be selected as a suitable constant in real-world applications. From (3.17), fractional-order state feedback is essentially a fractional-order differential of the hidden states $h_p(k)$. Before discretization, the fractional-order system presented by (3.14) still has a Mittag-Leffler stability problem, which would influence the embedding into the network and the training process of the proposed FORNN with PIBatKnow.

3.2.2 Fractional-order gradient descent method

Besides fractional-order state feedback and fractional-order constraint, FORNN can also be enhanced in the training backward process by fractional-order gradients, as shown in Figure 3.6.

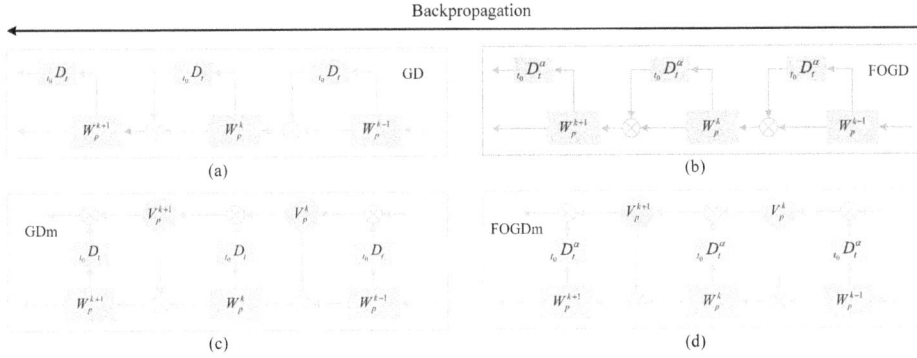

Figure 3.6 Fractional-order gradient descent methods. (a) integer-order GD method, (b) fractional-order GD method (FOGD), (c) integer-order GD method with momentum (GDm), (d) fractional-order GD method with momentum (FOGDm).

In the opposite direction of forward propagation, backward process starts from the output layer to input layer, mainly updating the weights W_p, W_{hp}, and the bias b_p, $(p = 1, 2, ..., l)$ in the gradient descent direction, which is called gradient descent (GD) method as shown in Figure 3.6 (a). It always makes the weights $W_p = W_{hp}$ in real applications, and the bias b_p holds the same update with weight W_p, thus only weight W_p is discussed in the following. During backward process with backpropagation, the layer state gradients are calculated, then weight gradients are based on the layer gradients to update weight W_p. The gradients of the pth hidden layer for the kth, $(k = 1, 2, ..., N)$ input sample are presented as

$$
\begin{cases}
\dfrac{\partial L(g(x), y)}{\partial h_p^k(x)} = W_{p+1}^{\prime k} \dfrac{\partial L(g(x), y)}{\partial a_{p+1}^k(x)} \\[3mm]
\dfrac{\partial L(g(x), y)}{\partial a_p^k(x)} = \dfrac{\partial L(g(x), y)}{\partial h_p^k(x)} \cdot g'(a_p^k(x)) = W_{p+1}^{\prime k} \dfrac{\partial L(g(x), y)}{\partial a_{p+1}^k(x)} \cdot g'(a_p^k(x)),
\end{cases}
\tag{3.18}
$$

where $\partial L(g(x), y)/\partial h_p^k(x)$ and $\partial L(g(x), y)/\partial a_p^k(x)$ are the gradients of the output $h_p^k(x)$ and the input $a_p^k(x)$ of the pth hidden layer, respectively. Since the backpropagation of gradient starts from the output layer, it can be assumed that the gradients $\partial L(g(x), y)/\partial h_t^k(x)$ and $\partial L(g(x), y)/\partial a_t^k(x)$ of the tth($t = p + 1, p + 2, ..., l, l + 1$) layer are already known. With layer gradients in (3.18), fractional-order derivative is introduced from the output layer to calculate the fractional-order gradients of loss function $L(g(x), y)$, then all the weights gradients of hidden layers are transferred to fractional-order ones, which turns out to be a fractional-order gradient descent

(FOGD) method as shown in Figure 3.6 (b). For time-series battery dataset as the inputs of the proposed FORNN, if considering gradient descent method with momentum (GDm) shown in Figure 3.6 (c), FOGD method can be extended to FOGD method with momentum (FOGDm) as illustrated in Figure 3.6 (d).

3.2.2.1 Fractional-order gradient descent

FOGD employs the fractional-order gradients of loss function and weights W_p instead of the integer-order gradients. Based on the fractional-order gradient in (3.9), FOGD for the updates of weight W_p is presented as

$$W_p^{k+1} = W_p^k - \eta \cdot_{W_p^0} D_{W_p^k}^\alpha L(g(x), y) = W_p^k - \eta \cdot \frac{\partial^\alpha L(g(x), y)}{\partial \left(W_p^k\right)^\alpha} \tag{3.19}$$

According to $a_p(x) = W_p h_{p-1}(x) + b_p$ and the approximate chain rule [18], (3.19) is deduced as

$$W_p^{k+1} \approx W_p^k - \eta \cdot \frac{\partial L(g(x), y)}{\partial a_p^k(x)} \cdot \frac{\partial^\alpha a_p^k(x)}{\partial \left(W_p^k\right)^\alpha} \tag{3.20}$$

where $\partial L(g(x), y)/\partial (W_p^k)^\alpha$ is the fractional-order gradients of weight W_p to the loss function $L(g(x), y)$, and η is the learning rate (iteration step size).

Remark 3.1: It should be noted that the chain rule used in (3.20) is an approximation result [18]. Using the generalized Leibniz series expansion, the fractional derivative of a composite function $f(g(x))$ should have an infinite term as

$$D^\alpha f(g(x)) = \sum_{k=0}^\infty \frac{\Gamma(1+\alpha)}{\Gamma(1+k)\Gamma(1+\alpha-k)} g^{(\alpha-k)}(x) f^{(k)}(g(x)) \tag{3.21}$$

where $g^{(\alpha-k)}(x)$ represents the fractional derivatives of the inner function $g(x)$. The chain rule for the fractional derivative of a composite function $f(g(x))$ can only obtain numerical approximation within a certain range of x, and would have large errors when x increases. Hence, the use of chain rule should be carefully considered.
Example: To illustrate the approximation effects of the chain rule used in (3.20), we provide an example here. Let's compute a fractional derivative of the composite function $f(x) = (x^2 + 1)^{3/2}$ with fractional order $\alpha = 1/2$ using both chain rule and numerical G-L method for comparison.

The composite function $f(x) = (x^2 + 1)^{3/2}$ has
1. Outer function: $f(y) = y^{3/2}$ and,
2. Inner function: $g(x) = x^2 + 1$.
Then, the first-order derivative of the outer function is

$$f'(y) = \tfrac{3}{2} y^{1/2}.$$

The first-order derivative of the inner function is $g'(x) = 2x$, and the fractional-order derivative ($\alpha = 1/2$) of $g(x)$ is

$$D^{1/2}g(x) = D^{1/2}\left(x^2 + 1\right) \approx \frac{\Gamma(3/2)}{\Gamma(2)}x^{1/2} = \frac{1/2 \cdot \sqrt{\pi}}{\Gamma(2)}x^{1/2} = \frac{\sqrt{\pi}}{2}x^{1/2}.$$

Apply the chain rule, the fractional-order ($\alpha = 1/2$) derivative of the composite function $f(g(x)) = (x^2 + 1)^{3/2}$ can be approximated as

$$\begin{aligned}
D^{1/2}f(g(x)) &\approx f'(g(x))D^{1/2}g(x) \\
&= \frac{3}{2}\left(x^2 + 1\right)^{1/2} \cdot \frac{\sqrt{\pi}}{2}x^{1/2} \\
&= \frac{3\sqrt{\pi}}{4}x^{1/2}\left(x^2 + 1\right)^{1/2}
\end{aligned}$$

Compare the approximation result by chain rule to the numerical result by G-L method, we can obtain the comparison as shown in Figure 3.7.

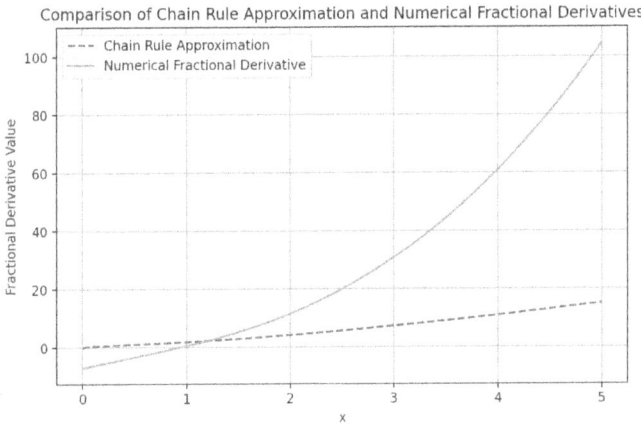

Figure 3.7 Comparison of the approximation result by chain rule to the numerical result by G-L method ($x \in (0, 5)$).

In (3.20), the gradient of the input $a_p^k(x)$ to the loss function ($\partial L(g(x), y)/\partial a_p^k(x)$) can be obtained by (3.18) and $\partial^\alpha a_p^k(x)/\partial(W_p^k)^\alpha$ can be calculated by (3.5). Hence, combining (3.18) and the discrete Caputo definition in (3.5), the fractional-order gradients of weight W_p to the loss function $L(g(x), y)$ can be deduced as

$$\frac{\partial^\alpha L(g(x), y)}{\partial(W_p^k)^\alpha} = \frac{h_{p-1}^k(x)}{\Gamma(2 - \alpha)} \cdot \frac{\partial L(g(x), y)}{\partial a_p^k(x)}\left(W_p^k - W_p^0\right)^{1-\alpha} \tag{3.22}$$

where W_p^0 is the initial values of weight W_p. Take (3.22) into (3.20), we can obtain the FOGD method with backpropagation as

$$W_p^{k+1} = W_p^k - \eta \cdot \frac{h_{p-1}^k(x)}{\Gamma(2-\alpha)} \cdot \frac{\partial L(f(x), y)}{\partial a_p^k(x)} \left(W_p^k - W_p^0\right)^{1-\alpha} \qquad (3.23)$$

where α is a fractional order that may be related to battery dynamics and sensitive to the training results. The sensitivity of the fractional order α and the corresponding influence on network output is discussed in the experimental part.

3.2.2.2 Fractional-order gradient descent with momentum

FOGDm method is extended from FOGD method by adding a momentum term, which demonstrates the changing direction of the fractional-order gradients. Given the integer-order gradient descent (GD) method with momentum as

$$\begin{cases} V_p^{k+1} = \mu \cdot V_p^k - \eta \cdot \frac{\partial L(g(x), y)}{\partial W_p^k} \\ W_p^{k+1} = W_p^k + V_p^{k+1} \end{cases} \qquad (3.24)$$

where μ is the momentum factor, which is proportional to history gradient data, and V_p is the weight momentum. Then the weight update by FOGDm method can be presented as

$$\begin{cases} V_p^{k+1} = \mu \cdot V_p^k - \eta \cdot \frac{\partial^\alpha L(g(x), y)}{\partial \left(W_p^k\right)^\alpha} \\ W_p^{k+1} = W_p^k + V_p^{k+1}. \end{cases} \qquad (3.25)$$

Taking the fractional-order gradients of weight W_p to the loss function $L(g(x), y)$ in (3.22) into (3.25), we obtain the discrete updating equation of weight W_p by FOGDm method as

$$\begin{cases} V_p^{k+1} = \mu \cdot V_p^k - \eta \cdot \frac{h_{p-1}^k(x)}{\Gamma(2-\alpha)} \cdot \frac{\partial L(g(x), y)}{\partial a_p^k(x)} \left(W_p^k - W_p^0\right)^{1-\alpha} \\ W_p^{k+1} = W_p^k + V_p^{k+1}. \end{cases} \qquad (3.26)$$

With (3.26) to encode the momentum of fractional-order gradient into the backward process with backpropagation, FOGDm method may accelerate the convergence of the proposed FORNN with faster updates of weight W_p and bias b_p.

3.2.3 Fractional-order constraint

As provided in the introduction Section 3.1.1, the embedded physics information in ML algorithms was mainly embodied in observational bias, inductive bias, and learning bias [8]. A typical example of learning bias is that physics is enforced via soft penalty constraints into the loss function of NNs. In this subsection, the soft penalty constraint is realized by a fractional-order PDE reflecting battery knowledge and embedded into the loss function of RNN, as shown in Figure 3.8. In addition to the common loss of the supervised measurement of outputs, a fractional-order PDE driven by physics-informed knowledge is included as an unsupervised fractional-order loss term, to form the final total loss for training the backward process.

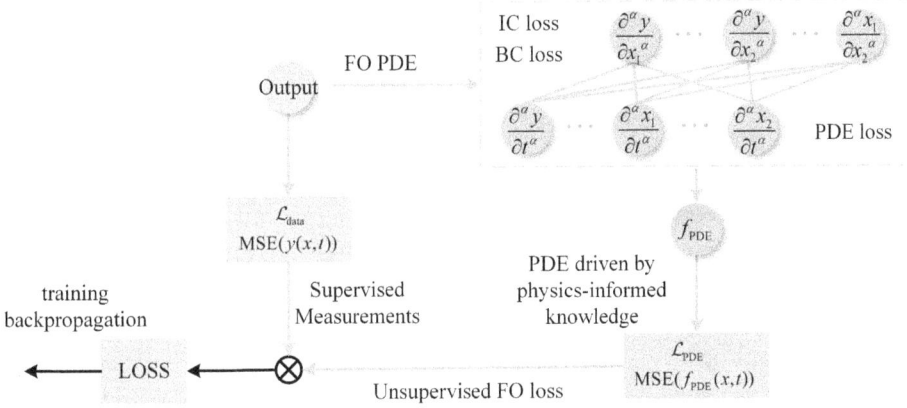

Figure 3.8 Fractional-order partial differential equation constraint.

3.2.3.1 Constraint in fractional-order PDE form

As shown in Figure 3.8, the unsupervised constraint is in a fractional-order PDE form, which can include various types of fractional-order derivatives, such as PDE loss, initial condition (IC) loss, and boundary condition (BC) loss. The fractional-order PDE constraint is first presented in this part, then the specific fractional-order PDE for LIB is presented in the following part. The general form of a PDE with initial condition and boundary condition is given as follows.

PDE with initial and boundary conditions [17].

$$
\begin{aligned}
\mathcal{L}[y](x) &= f(x), && x \in \Omega \times [0, T], \\
\mathcal{B}[y](x) &= y_b(x), && x \in \partial\Omega \times [0, T], \\
y(x, 0) &= y_0(x), && x \in \Omega,
\end{aligned}
\tag{3.27}
$$

where $\mathcal{L}$ and $\mathcal{B}$ represent the differential operators, Ω is the spatio-temporal domain of interest, y is the solution, and y_b and y_0 are the boundary and initial functions.

The exact solution y^* of the PDE usually lies in an infinite dimensional space, and the FORNN with PDE constraint in this chapter would parameterize the solution y^* of as y (3.27) to approximate the ground truth in a numerical way. If the differential operator in (3.27) is chosen as integer-order derivative, it can obtain an integer-order PDE constraint as stated in the follows.

If the differential operators in (3.27) are integer-order, a simple physics-informed PDE constraint can be presented as

$$
f_{\text{PDE}}(x, t) = \frac{\partial y}{\partial t} + a_1 \frac{\partial y}{\partial x} + a_2 \frac{\partial^2 y}{\partial x^2},
\tag{3.28}
$$

where a_1 and a_2 are the coefficients of boundary condition loss (BC loss) and initial condition loss (IC loss). The PDE constraint in (3.28) satisfies certain physics-informed law of the object, and (3.28) can be extended to a complex form with extra information by adding other types of derivatives more than just IC loss or BC loss.

Remark 3.2: The PDE constraint in (3.28) could be integer-order PDEs, integro-differential equations, fractional-order PDEs or stochastic PDEs [8].

As stated in **Remark 3.2**, the PDE constraint in (3.28) may vary from different physical equations according to the investigated object in different problems, such as the viscous Burgers' equation $\frac{\partial y}{\partial t} + y\frac{\partial y}{\partial x} = v\frac{\partial^2 y}{\partial x^2}$ with an initial condition and Dirichlet boundary conditions. Considering the embedding battery knowledge represented by FOMs in this chapter, the integer-order PDE constraint in (3.28) could be extended to a fractional-order one as stated in the follows. Combining with the various forms of fractional-order derivatives in Figure 3.8, and if the differential operators in (3.27) is fractional order, the fractional-order PDE constraint with physics-informed knowledge can be represented as

$$f_{\text{FOPDE}}(x,t) = \underbrace{\frac{\partial^\alpha y}{\partial t^\alpha}}_{\text{PDE loss}} + \underbrace{a_1\frac{\partial^\alpha y}{\partial x_1^\alpha} + a_2\frac{\partial^\alpha y}{\partial x_2^\alpha}}_{\text{condition loss}} + \underbrace{b_1\frac{\partial^\alpha x_1}{\partial t^\alpha} + b_2\frac{\partial^\alpha x_2}{\partial t^\alpha}}_{\text{input constraint}} + \underbrace{c_1\frac{\partial^\alpha x_1}{\partial x_2^\alpha}}_{\text{coupled rate}}, \quad (3.29)$$

where $a_i, i = 1, 2$, $b_i, i = 1, 2$, and c_1 are the coefficients of condition loss (such as IC loss and BC loss), input constraint, and coupled rate, respectively. The fractional-order PDE constraint $f_{\text{FOPDE}}(x,t)$ in (3.29) normalizes the coefficient of PDE loss ($\partial^\alpha y/\partial t^\alpha$) as 1, and sets the fractional order α as the same, which could be different in some applications. The fractional-order PDE constraint $f_{\text{FOPDE}}(x,t)$ in (3.29) satisfies a certain fractional-order physics-informed law of the object.

The fractional-order PDE constraint in (3.29) contains PDE loss, initial boundary condition loss, input constraint, and coupled constraint. Among the four parts, input constraint may be the input derivative in time series, and coupled constraint may be the coupled rate or coupled relationships among inputs, which usually exists in physical systems. Note that the fractional-order PDE in (3.29) covers the integer-order one in (3.28) when the fractional-order α is set as 1.

According to [8], and referring to the fractional-order PDE constraint in (3.29), the final loss function can be deduced as

$$\mathcal{L} = w_{\text{data}}\mathcal{L}_{\text{data}} + w_{\text{PDE}}\mathcal{L}_{\text{PDE}}, \quad (3.30)$$

where

$$\mathcal{L}_{\text{data}} = \frac{1}{N_{\text{data}}}\sum_{i=1}^{N_{\text{data}}}(y(x_i, t_i) - y_i)^2 \quad (3.31)$$

is the supervised loss of data output error, and

$$\begin{aligned}\mathcal{L}_{\text{PDE}} &= \frac{1}{N_{\text{PDE}}}\sum_{j=1}^{N_{\text{PDE}}}[f_{\text{FOPDE}}(x,t)]^2\big|_{(x_j,t_j)} \\ &= \frac{1}{N_{\text{PDE}}}\sum_{j=1}^{N_{\text{PDE}}}\left(\frac{\partial^\alpha y}{\partial t^\alpha} + a_i\frac{\partial^\alpha y}{\partial x_m^\alpha} + b_i\frac{\partial^\alpha x_m}{\partial t^\alpha} + c_i\frac{\partial^\alpha x_m}{\partial x_n^\alpha}\right)^2\big|_{(x_j,t_j,m=1,2,m\neq n,n=1,2)}\end{aligned} \quad (3.32)$$

is the unsupervised loss of fractional-order PDE constraint $f_{\text{FOPDE}}(x, t)$.

In (3.30) to (3.32), w_{data} and w_{PDE} are the weights to balance the two losses $\mathcal{L}_{\text{data}}$ and $\mathcal{L}_{\text{PDE}}$; y_i are values of y at (x_i, t_i); (x_i, t_i) and (x_j, t_j) are two sets of points sampled at the initial/boundary locations and in the entire domain, respectively.

With (3.29) and (3.30), the general form of the fractional-order PDE constraint is proposed to encode physics knowledge into the loss function of FORNN. For specific object, the details of the physics-informed PDE in (3.29) should be extracted from equations of the mechanism modeling for this object.

3.2.3.2 Constraint of battery terminal voltage derivative equation

As introduced in Section 3.1.2, the physics laws of LIBs can be described by fractional-order elements in battery FOM, which is a suitable choice to be transferred to the fractional-order PDE constraint as stated in (3.29). A second-order FOM is first considered in this chapter as shown in Figure 3.9 [3, 4].

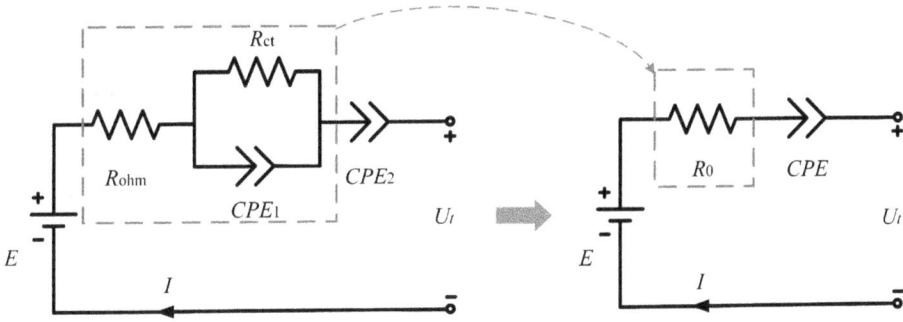

Figure 3.9 Simplified FOM of LIBs as fractional-order PDE constraint.

In Figure 3.9, the impedance of the second-order FOM in the left side can be presented as

$$Z = R_{\text{ohm}} + Z_{\text{ARC}} + Z_{\text{Warburg}} = R_{\text{ohm}} + \frac{R_{ct}}{R_{ct}C_1 s^{\alpha_1} + 1} + \frac{1}{C_2 s^{\alpha_2}} \qquad (3.33)$$

where E is the open circuit voltage (OCV), U_t is the terminal voltage, and R_{ohm}, RC tank Z_{ARC} (R_{ct} and CPE_1), and Z_{Warburg} (CPE_2) represent the ohmic resistance in high-frequency, double layer effects and charge transfer reactions in mid-frequency, and solid diffusion dynamics in low-frequency of EIS of LIBs, respectively. Specifically, the time constant of R_{ohm} and RC tank Z_{ARC} is less than 0.05s and Z_{Warburg} is longer than 50s in the time scale [3], while the sampling frequency of general BMS is about $1 \sim 10$ Hz, which cannot accurately identify the transient response of R_{ohm} and the dynamic response of Z_{ARC}. Considering R_{ohm} and RC tank Z_{ARC} as one process, the second-order FOM is simplified as an RC tank as shown in the right side of Figure 3.9, which has the impedance as

$$\frac{U_t(s) - E(s)}{I(s)} = R_0 + Z_{\text{Warburg}} = R_0 + \frac{1}{C s^{\alpha}}, \qquad (3.34)$$

where $R_0 = R_{\text{ohm}} + Z_{\text{ARC}}$. Equation (3.34) acts as the basis of the fractional-order constraints for the loss function of PIRNN, which is further deduced later.

Using the simplified FOM in Figure 3.9, and the corresponding voltage-current equation (3.34) can be presented as

$$U_t(s) - E(s) = (R_0 + \frac{1}{Cs^\alpha})I(s)$$
$$\Rightarrow Cs^\alpha(U_t(s) - E(s)) = (R_0Cs^\alpha + 1)I(s), \tag{3.35}$$

where U_t is the terminal voltage, E is the OCV, and I is the current; R_0 and C are the resistance part and capacity value of the FOM in Figure 3.9, respectively.

Applying inverse Laplace transform, (3.35) in time domain becomes

$$C \cdot D^\alpha(u_t(t) - E(t)) = (R_0C \cdot D^\alpha + 1)i(t), \tag{3.36}$$

where $u_t(t)$, $E(t)$, and $i(t)$ are the time histories of the terminal voltage, OCV, and current in time domain, respectively. Rewrite (3.36) as

$$C\frac{\partial^\alpha u_t(t)}{\partial t^\alpha} = C\frac{\partial^\alpha E(t)}{\partial t^\alpha} + R_0C\frac{\partial^\alpha i(t)}{\partial t^\alpha} + i(t), \tag{3.37}$$

where the terminal voltage $u_t(t)$ and current $i(t)$ can be directly obtained from the realistic sampled EV data, while OCV $E(t)$ is an internal variable of battery that can be hardly measured in real operation conditions. However, SOC does have a relationship with OCV, which can be approximately expressed as [22]

$$E(t) = f(SOC) = \sum_{k=0}^{\infty} d_k SOC^k(t), \tag{3.38}$$

where d_k is the coefficients of $SOC^k(t)$. Considering (3.37) as the fractional-order PDE constraint and a finite terms k in (3.38), some assumptions are proposed to replace OCV in (3.37) by SOC, as stated in **Lemma 6**.

According to the measured SOC-OCV curves in the literature, let k in (3.38) hold $k = 1$ and assume OCV $E(t)$ is monotonic with $SOC(t)$, so that the fractional-order derivative of $E(t)$ holds the relationship with the fractional-order derivative of $SOC(t)$ as

$$\frac{\partial^\alpha E(t)}{\partial t^\alpha} = d_{soc}\frac{\partial^\alpha SOC(t)}{\partial t^\alpha}, \tag{3.39}$$

where d_{soc} is the ratio of $(\partial^\alpha E(t)/\partial t^\alpha)$ to $(\partial^\alpha SOC(t)/\partial t^\alpha)$.

Substituting (3.39) into (3.37), one can obtain the fractional-order PDE constraint from the FOM of battery as

$$C\frac{\partial^\alpha u_t(t)}{\partial t^\alpha} = d_{soc}C\frac{\partial^\alpha SOC(t)}{\partial t^\alpha} + R_0C\frac{\partial^\alpha i(t)}{\partial t^\alpha} + i(t). \tag{3.40}$$

For FORNN with PIBatKnow, the inputs and outputs should be carefully selected and combined with the fractional-order constraint (3.40). Moreover, it should consider the available variables from realistic data, the influence of temperature, and the aim of the proposed FORNN. For example, taking state estimation as the target, then the inputs are defined as $\mathbf{x}(t) = [x_1(t), x_2(t), x_3(t)]^T = [u_t(t), i(t), T_{temp}(t)]^T$, and the output as $y(t) = SOC(t)$, thus the fractional-order PDE constraint can be written as

$$f_{SOC}(x,t) = d_{soc}C\frac{\partial^\alpha y(t)}{\partial t^\alpha} + R_0C\frac{\partial^\alpha x_2(t)}{\partial t^\alpha} + x_2(t) - C\frac{\partial^\alpha x_1(t)}{\partial t^\alpha}. \qquad (3.41)$$

Combining (3.41) with (3.32), the loss of fractional-order PDE constraint can be calculated in the form of mean sum of squares (MSS) and embedded into the loss function as a fractional-order unsupervised loss. G-L definition can be employed to discretize the fractional-order derivatives in (3.41) for numerical calculation. Select the number of the finite terms $L = 5$ in (3.2), and the fractional-order derivatives of inputs $x(t)$ and output $y(t)$ in (3.41) can be approximated as

$$\Delta^\alpha M(t) \doteq \sum_{j=0}^{5} (-1)^j \begin{pmatrix} \alpha \\ j \end{pmatrix} M(k-j), \qquad (3.42)$$

where $M(t) = x_1(t), x_2(t),$ or $y(t)$, and k is the discrete step. Note that the discrete equation in (3.42) requires the history data in previous $\{5 - j, (j = 0, 1, ..., 5)\}$ moments, which can be found in the time-series data of the realistic sampled voltage, current, and the output SOC. With (3.41) and (3.42), an FORNN with a fractional-order PDE constraint is constructed and encoded by battery voltage equation, then the proposed FORNN can be applied to conduct SOC estimation or further life prediction.

3.2.4 Framework and training procedure

With the FOGDm method in Section 3.2.2 and fractional-order PDE constraints in Section 3.2.3, we can construct a PIRNN with fractional-order constraints, as shown in Figure 3.10, named **fPIRNN**. The traditional RNN acts as the basic structure

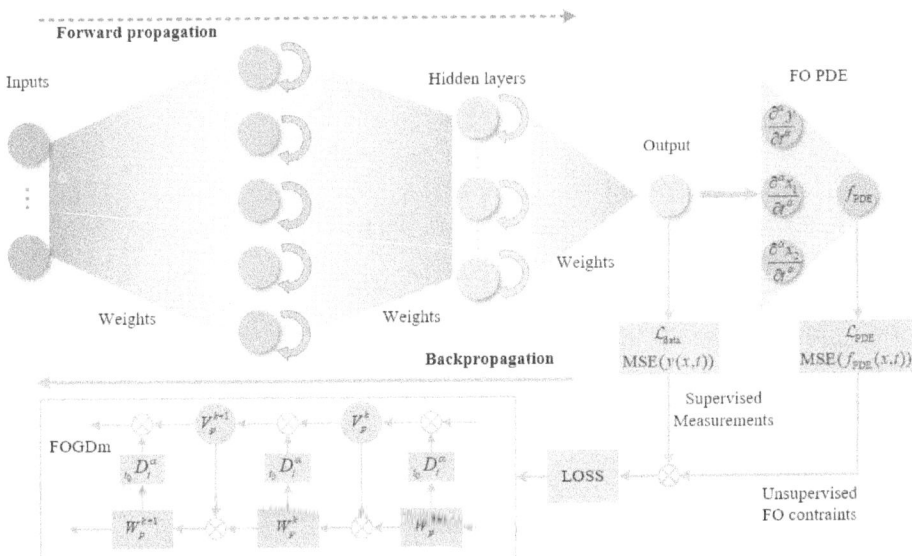

Figure 3.10 Framework of the proposed PIRNN with fractional-order constraints.

of our fPIRNN, which still maintains the time-series state feedback and the chain structure in the forward propagation, while the backward process is revised extensively by backpropagation of FOGDm in (3.23) and fractional-order PDE constraints in (3.30)–(3.32) and (3.41). In every epoch, unsupervised fractional-order constraints $\mathcal{L}_{\text{PDE}}$ are calculated by the fractional-order PDE following the output layer; then, by combining unsupervised loss $\mathcal{L}_{\text{PDE}}$ with supervised measuring loss $\mathcal{L}_{\text{data}}$, fPIRNN can update weights with the final loss and the FOGDm method. The physics-informed PDE is embedded in unsupervised loss $\mathcal{L}_{\text{PDE}}$ to instruct the backward process of fPIRNN and FOGDm is also employed to accelerate the convergence of the network. From Figure 3.10, the implementation of fPIRNN should discretize the state feedback of chain structure and the fractional-order gradients of FOGDm method. We list the details of the training steps in Table 3.1.

Table 3.1 Training procedure of the PIRNN with fractional-order constraints (fPIRNN)

Procedure 3.1 Training details of fPIRNN for SOC estimation of LIBs	
input:	Sampled dataset $D = \{(x_{1,1}, x_{1,2}, x_{1,3}, y_1), (x_{2,1}, x_{2,2}, x_{2,3}, y_2), \ldots, (x_{n,1}, x_{n,2}, x_{n,3}, y_n)\}$
step 1	Construct a basic RNN with specific hidden layers, neurons, and state feedbacks.
step 2	Divide dataset into training dataset, validation dataset, and testing dataset.
step 3	Initialization. Specify the initial parameters values of fPIRNN, weight W_p^0, epoch threshold E_{max}, learning rate η, fractional order α, desired loss $\mathcal{L}_{\text{target}}$.
step 4	While epoch $\leq$ threshold $Epoch_{\text{max}}$: Forward propagation, calculate neuron states from the input layer to the output layer. Calculate the measured loss of the output $\mathcal{L}_{\text{data}}$. Calculate the PDE loss $\mathcal{L}_{\text{PDE}}$ under the FO constraints by (3.31), (3.39), and (3.40). Calculate training loss $\mathcal{L}$ with $\mathcal{L}_{\text{data}}$ and $\mathcal{L}_{\text{PDE}}$. Backpropagation, update weights W_p with the FOGDm method by (3.23). epoch = epoch +1
step 5	If $\mathcal{L}$ achieves $\mathcal{L}_{\text{target}}$: training finished; else: adjust setup and back to step 2 to re-train.
output:	Trained fPIRNN with converged parameters, which can make predictions on the testing dataset.

3.3   STATE OF CHARGE ESTIMATION

To verify the proposed fPIRNN outlined in Figure 3.10, we have conducted experiments under FUDS operation conditions and sampled necessary battery data. Parameter sensitivity was first analyzed and then fPIRNN was trained with the sampled FUDS data in accordance with the procedure stated in Table 3.1. SOC estimation results under FUDS are presented, and corresponding discussions are also provided in this section.

3.3.1   Experiment setup

Four 18650 cells are considered as test objects and numbered as cell1, cell2, cell3, and cell4, and detailed parameters are listed in Table 3.2. The experiment setup is shown in Figure 3.11. FUDS operation condition is applied to the four 18650 cells from the BTS-4 series battery tester produced by Neware Company to simulate a realistic working situation, and an incubator is included to maintain the environment temperature of the experiments. LIB cells are fully charged by the constant-current constant-voltage (CC-CV) method before applying FUDS; then, the cells are fully discharged to the cut-off voltage of 2.75 V in Table 3.2 under cycling FUDS periods (a single period in FUDS shown in Figure 3.12), and data are collected by the BTS-4 tester and sent to a host computer for further processing.

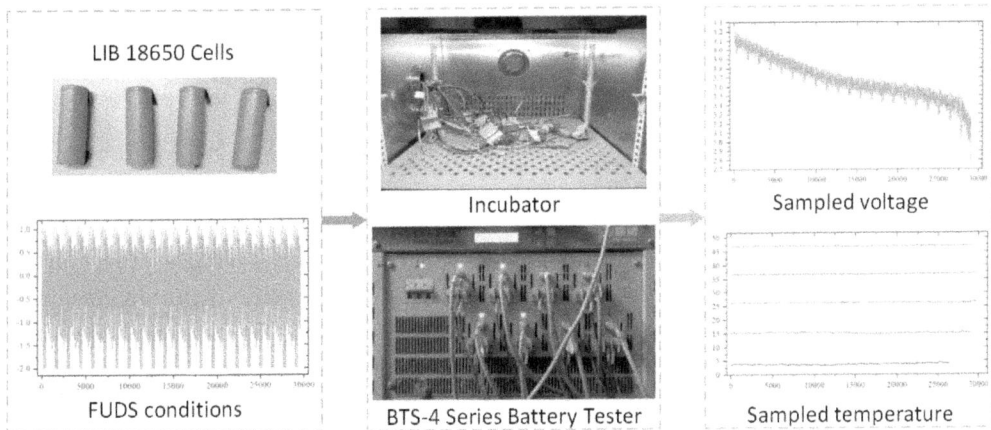

Figure 3.11   Experiment setup for LIB tests.

Considering the available capacity of battery cells in realistic working conditions, temperatures induce significant influences on the discharging/charging ability of battery cells from 0 to 100% SOC range, which may cause different SOC-changing trajectories. Hence, experiments under five temperatures were conducted, and the available capacity of the four cells is measured, as shown in Figure 3.13. Real SOC values are calculated by the ampere-hour integral method with the capacity listed in Figure 3.13. It can be found in Figure 3.13 that the influence of temperature $T_{temp}(t)$ cannot be ignored; thus, as presented in the previous section, the proposed fPIRNN determines

Table 3.2 LIB 18650 cell parameters

Parameters	Values
rated capacity $(0.5C_5A)$	2000 mAh
rated voltage	3.7 V
max charge voltage	4.2 V
discharge cut-off voltage	2.75 V
max charging current	$1C_5$ A
max discharging current	$2C_5$ A

the inputs as $\mathbf{x}(t) = [x_1(t), x_2(t), x_3(t)]^T = [u_t(t), i(t), T_{temp}(t)]^T$ and the output as $y(t) = SOC(t)$.

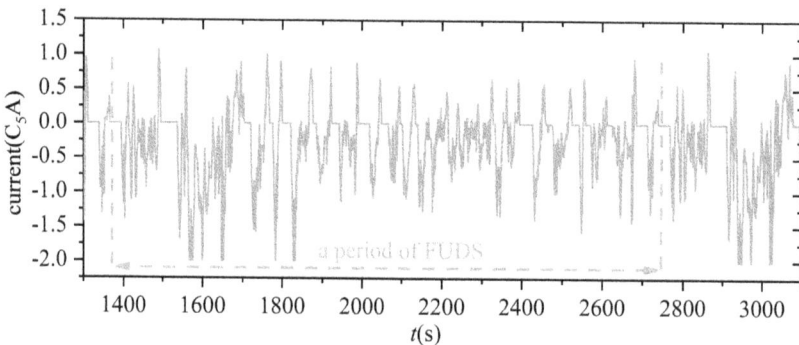

Figure 3.12 Current profile of a single period in FUDS.

The data for the proposed fPIRNN are divided into training, validation, and testing data. The data in this study are sampled by the BTS-4 battery tester in 1 Hz frequency. With preprocessing, the total amount of data is 27213 points, which can support the big-data learning during training process for the proposed algorithm. The collected data is divided according to the five temperatures, that is, 45°C is selected as the testing data, and data under 5°C, 15°C, 25°C, and 35°C is divided into training data and validation data. The division ratio is 0.75 : 0.048 : 0.202, resulting in training data (20410 points), validation data (1306 points), and testing data (5497 points).

3.3.2 Sensitivity analysis

The proposed FORNN with PIBatKnow in this chapter may contain fractional-order state feedback, fractional-order loss constraint, and fractional-order gradient with backpropagation, which bring more parameters for algorithm tuning. Hence, the sensitivity analysis is necessary to identify suitable values and maximize the effectiveness for the practical estimation of LIB. Since the aim of this work is sensitivity analysis of parameters, only the data of one cell is enough, and any one of five temperatures (not specific 45°C) can be selected as the testing data. Moreover, some unchanged

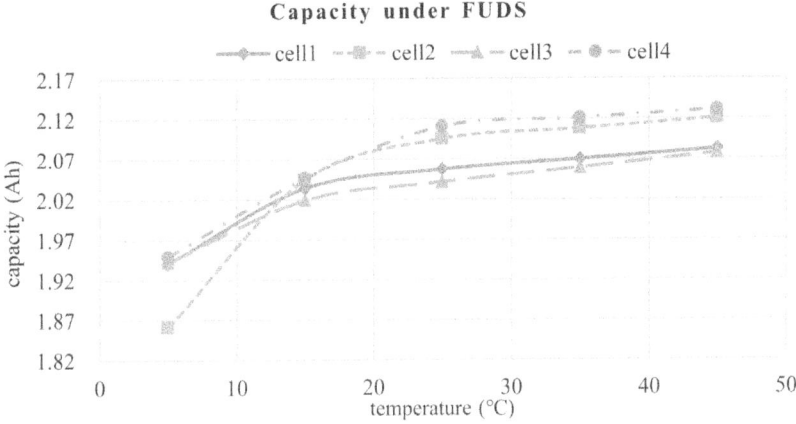

Figure 3.13 The measured capacity values of four 18650 cells under five temperatures (5°C,15°C, 25°C, 35°C, and 45°C).

parameters are initialized and listed in Table 3.3, and the performance is calculated by mean squared error (MSE).

Table 3.3 Unchanged parameters of the proposed FORNN with PIBatKnow

Name	Value/Range	Name	Value/Range
hidden layers	1	hidden neurons	12
max epoch E_{max}	300	performance function	MSE
train : valid : test	0.75: 0.048: 0.202	training goal	1.6e-4

Nine fractional-order parameters are investigated and divided into three categories as shown in Table 3.4, that is, parameters sensitivity in FOGD and FOGDm method (fractional-order gradient sensitivity), parameters sensitivity in fractional-order PDE constraint (impedance sensitivity), and weights in loss calculation (loss weight sensitivity). The range of the parameters are also provided in Table 3.4, in which only capacitance C_{bat} and resistance R_0 have physical units, that is, F and Ω, respectively, and other seven fractional-order parameters are unitless.

3.3.2.1 Estimation with fractional-order gradient sensitivity

The first sensitivity category is the fractional-order gradient sensitivity, which contains the fractional order α_1, the momentum weight μ, and the learning rate η in FOGD and FOGDm method. Fractional order reflects and embeds battery fractional-order characteristics. In this section, only fractional order α_1, momentum weight μ, and learning rate η change in the range provided in Table 3.4, respectively. Other parameters stay unchanged as shown in Table 3.5, and the default values of fractional order α_1, momentum weight μ, and learning rate η are also provided when acting as the unchanged values in the other two sensitivity categories.

Table 3.4 Sensitivity categories of the fractional-order constraint and the fractional-order gradient in the proposed algorithm

Type	Name	Value/Range	Attribution
Fractional-order gradient sensitivity	fractional order α_1	$[0.1, 1]$	FOGDm in (3.26)
	momentum weight μ	$[0.1, 1]$	
	learning rate η	$[0.08, 0.26]$	
Impedance sensitivity	fractional order α_2	$[0.1, 1]$	$f_{SOC}(x, t)$ in (3.41)
	ratio of OCV-SOC d_{soc}	$[5, 50]$	
	capacitance C_{bat} (unit: F)	$[2.5, 25]$	
	ohm resistance R_0 (unit:Ω)	$[5e\text{-}3, 6e\text{-}3]$	
Loss weight sensitivity	loss weight w_{data}	$[0.1, 1]$	final loss $\mathcal{L}$ in (3.29)
	loss weight w_{PDE}	$1 - w_{data}$	

Table 3.5 Default values of the nine main fractional-order parameters

α_1	μ	η	α_2	d_{soc}	C_{bat}	R_0	w_{data}	w_{PDE}
0.9	0.75	0.18	0.9	40	20	0.005	0.8	0.2

To verify the estimation effects, SOC is taken as the output of the proposed FORNN with PIBatKnow. Figure 3.14 presents the SOC estimation results under the sensitivity of the fractional order α_1 ($\alpha_1 = 0.1 : 0.1 : 1$) in FOGD and FOGDm methods. Expression $variable = a : interval : b$ denotes that $variable$ changes from a to b in the range $[a, b]$ and takes the values in every interval $interval$, then this expression is applied to all the parameter sensitivity in the following. Figure 3.14 (b) contains five discharging snippets under FUDS condition in five temperatures (5°C, 15°C, 25°C, 35°C, 45°C). Snippet in 45°C is the testing dataset, whose outputs and errors are enlarged in Figure 3.14 (c) and Figure 3.14 (e), respectively.

The testing loss in Figure 3.14 (a) shows that loss decreases and converges faster with larger α_1, but the training process turns unstable when $\alpha > 0.5$, especially so when $\alpha = 1$. Moreover, the outputs of all dataset (training, validation, and testing) and errors (errors=target-outputs) are presented in Figure 3.14 (b) and (d), respectively. Figures 3.14 (b) and (d) also show the unstable accuracy performance. For the fractional order α_1, there exists a trade-off between performance and stability. With the outputs and the corresponding errors, it proves that the training dataset and testing dataset have similar performance, and the network does not overfit nor underfit. Similarly, the sensitivity of momentum μ and learning rate η are conducted in the same way as in Figure 3.14. Hence, the details of momentum μ and learning rate η are not presented repeatedly.

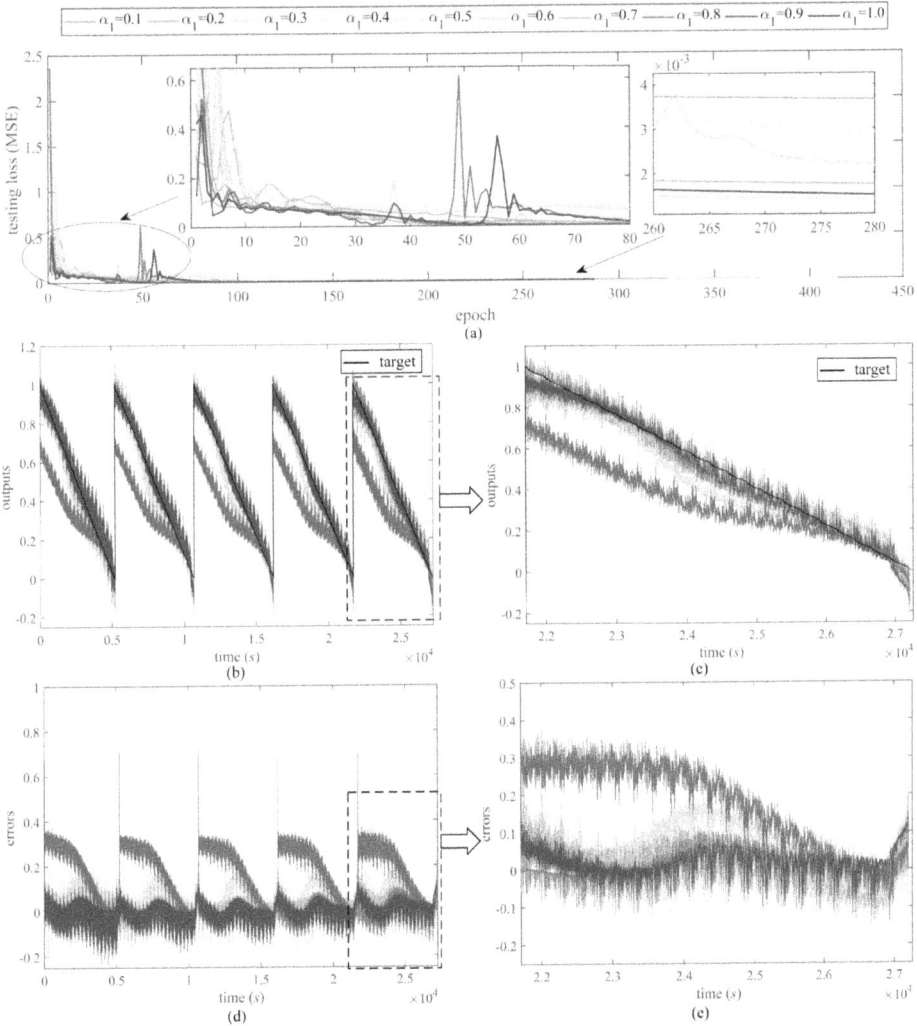

Figure 3.14 Sensitivity of the fractional order α_1 in FOGD and FOGDm methods. $\alpha_1 = 0.1 : 0.1 : 1$, (a) testing loss, (b) outputs of all dataset including training, validation, and testing data, (c) outputs of testing dataset, (d) errors of all dataset, (e) errors of testing dataset.

3.3.2.2 Estimation with impedance sensitivity

The second sensitivity category is the impedance sensitivity. For the fractional-order PDE constraint in loss function of the proposed FORNN with PIBatKnow, the main parameters include α_2, d_{soc}, C_{bat}, and R_0 of the battery FOM, and the sensitivity of the four impedance parameters are investigated in this section.

Figure 3.15 shows the testing loss, testing outputs, and corresponding errors under the impedance parameters α_2. To better illustrate the loss value, the testing loss is enlarged within the early stage of epochs, and a subfigure enlarging the stable loss in later epochs is added in every figure of testing loss, like Figure 3.15(a). Part of the

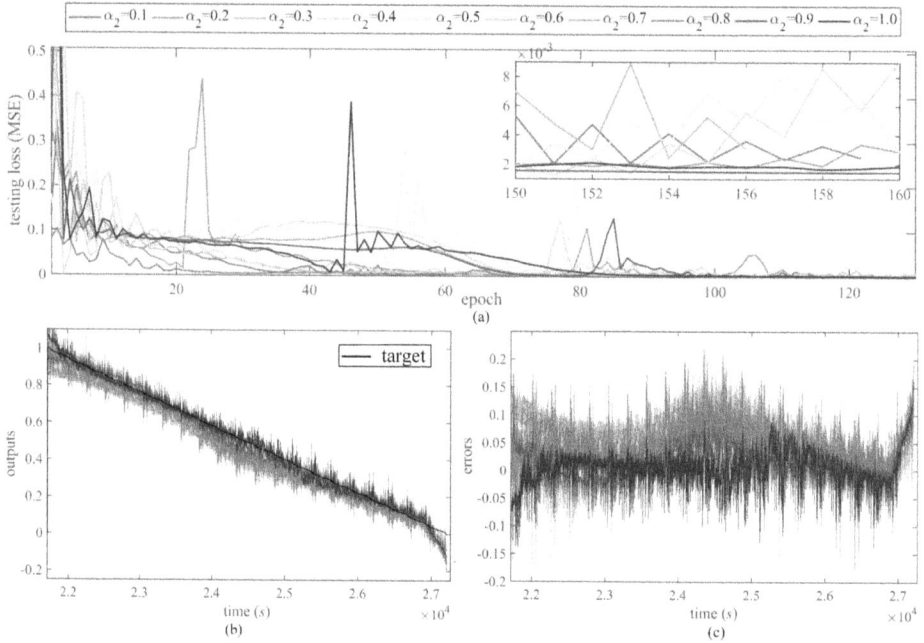

Figure 3.15 Sensitivity of the fractional order α_2 in PDE constraint encoded into loss function, $\alpha_2 = 0.1 : 0.1 : 1$. (a) testing loss, (b) testing outputs, (c) testing errors.

training process would be terminated before the max epoch (300) when reaching the training goal (1.6e-4) or other validation conditions. For example, in the subfigure Figure 3.15 (a), the training process with $\alpha_2 = 0.1$ stopped at the 159th epoch. From the accuracy aspect, the outputs and errors in Figure 3.15 do not show large fluctuations when the impedance parameters change. The results of d_{soc}, C_{bat}, and R_0 are similar to that of α_2, thus it is not necessary to show the figures repeatedly.

3.3.2.3 Estimation with loss weight sensitivity

The third sensitivity category is the loss weight sensitivity. The loss weight w_{data} and w_{PDE} can determine the ratio of the supervised measurement loss and the unsupervised fractional-order loss calculated by physics-informed PDE in Figure 3.8. Since loss weight w_{data} and w_{PDE} hold the relationship $w_{\text{PDE}} = 1 - w_{\text{data}}$, the two parameters may be considered as one parameter sensitivity, as presented in Figure 3.16. $w_{\text{data}} = 0.1 : 0.1 : 1$ corresponds to $w_{\text{PDE}} = 0.9 : 0.1 : 0$. The testing loss in Figure 3.16 (a) demonstrates that larger w_{data} makes the network process converge faster and more stable, which reflects the relative instability of the physics-informed knowledge embedding. However, the algorithm can achieve a trade-off with sensitivity analysis and suitable tuning. As to the estimation accuracy, the outputs and errors in Figure 3.16 (b) and Figure 3.16 (c) do not show much improvement both in small value and large value of w_{data}.

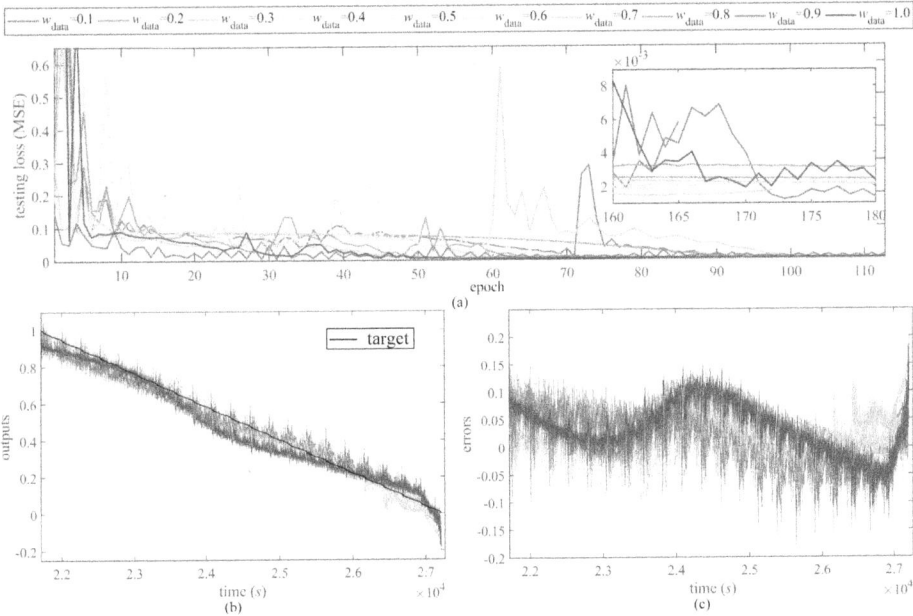

Figure 3.16 Sensitivity of the loss weights w_{data} and w_{PDE} in loss function calculation, $w_{\text{data}} = 0.1 : 0.1 : 1$. (a) testing loss, (b) testing outputs, (c) testing errors.

3.3.2.4 Sensitivity analysis conclusion and discussion

Combined with the estimation results under the three sensitivity categories, the sensitivity and correlation of the nine main fractional-order parameters to the algorithm performance can be concluded as shown in Table 3.6.

Table 3.6 Positive and negative correlation of the nine main fractional-order parameters to the algorithm performance

index	α_1	μ	η	α_2	d_{soc}	C_{bat}	R_0	w_{data}	w_{PDE}
speed[1]	↗	↘	↘	↘	↗	middle[4]	↘	↗	↘
loss[2]	↗	↗	-	↗	-	-	↘	-	-
accuracy[3]	↗	↗	-	↗	-	↗	↗	-	-
stability	↘	↘	↘	↗	↘	↗	↘	↗	↘

[1] speed means convergence speed. [2] loss means testing loss. [3] accuracy means estimation accuracy. [4] middle means that the algorithm achieves faster convergence speed in the middle values of C_{bat}.

Taking convergence speed, testing loss, estimation accuracy, and stability as the algorithm performance indicators, we can observe and present several notes listed below as references and instructions for the tuning of the nine main fractional-order parameters in the proposed FORNN with PIBatKnow.

- For the convergence speed, it could be boosted by a larger value of α_1, d_{soc}, and w_{data}, and by a smaller value of μ, η, α_2, R_0, and w_{PDE};

- For the testing loss, it could be improved by a larger value of α_1, μ, and α_2, and by a smaller value of R_0, but does not show sensitivity to η, d_{soc}, C_{bat}, w_{PDE}, and w_{data};

- For the estimation accuracy, it could be improved by a larger value of α_1, μ, α_2, C_{bat}, and R_0, but does not show sensitivity to η, d_{soc}, w_{PDE}, and w_{data};

- For the algorithm stability, it could be enhanced by a larger value of α_2, C_{bat}, and w_{PDE}, and by a smaller value of α_1, μ, η, d_{soc}, R_0 and w_{PDE};

- For the fractional order α_1 in FOGD and FOGDm methods, any value in the range $(0.5, 1)$ is suitable and a trade-off could be made between performance and stability;

- As the ratio of the previous momentum V_p, larger μ means larger inertia of the previous iteration, which makes the speed slow but improves the learning ability to achieve higher accuracy;

- The proposed algorithm achieves faster convergence speed in the middle values of the capacitance C_{bat} in battery FOM;

- The loss weight w_{PDE} has opposite tuning direction to the loss weight w_{data}.

3.3.3 Estimation results and comparison

3.3.3.1 Estimation with fractional-order constraints only

With preprocessed data and tuned fPIRNN, SOC estimations were performed under several conditions. The iterative method is chosen as the gradient descent with momentum (GDm) method. So an RNN with GDm is used as the benchmark for fair comparison purposes.

In Figure 3.17, the experiment results of fPIRNN with fractional-order PDE constraints and RNN with GDm are presented and marked as "FO constraints" and "GDm", respectively. A fitting process is employed for the outputs in this study by a second-order polynomial fitting for improved filtering and the presentation of estimation results (marked as "fitted output"). The corresponding output error is also calculated and marked as "fitted error", as shown in Figure 3.17(c). The relationships among the variables in Figure 3.17 and the following figures can be presented as follows:

$$
\begin{aligned}
error &= output - target, \\
fitted\ output &= \text{polyfit}(output), \\
fitted\ error &= fitted\ output - target,
\end{aligned}
\tag{3.43}
$$

where $output$ is the output SOC values, $target$ is the SOC target, and "polyfit" means the second-order polynomial fitting function.

Three main aspects of performance were analyzed, that is, convergence pro cess (Figure 3.17 a), output portraits (Figure 3.17 b,c), and estimation accuracy (Figure 3.17 d–f). Although both "FO constraints" and "GDm" can converge near

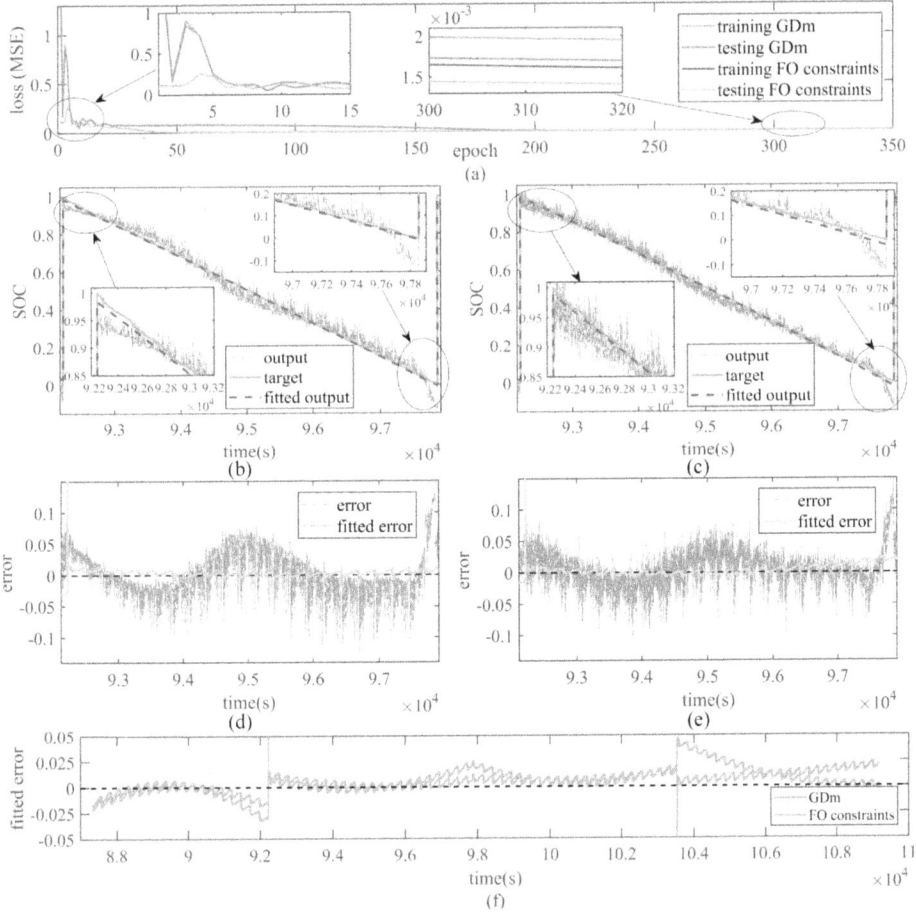

Figure 3.17 Experiment results of fPIRNN with fractional-order PDE constraints and RNN with GDm. (**a**) training and testing loss (marked as "training GDm", "testing GDm", "training FO constraints", and "testing FO constraints"), (**b**) estimated SOC with GDm, (**c**) estimated SOC with FO constraints, (**d**) estimated error with GDm, (**e**) estimated error with FO constraints, and (**f**) comparison of fitted errors.

$\mathcal{L}_{\text{target}}$ within 200 epochs, the convergence speed and achievable loss of fPIRNN with fractional-order constraints are superior to those of RNN with GDm. Under the disturbance of the switching point between two snippets, which have been preprocessed into a time-series sequence, the proposed algorithm can still output the estimated results stably, and the fitted output can follow the target during the entire SOC range (0–100%), as shown in Figure 3.17(c). With respect to the accuracy, less prediction noise in fPIRNN with fractional-order constraints was observed when comparing Figure 3.17(d),(e). Moreover, it can be observed that the proposed fPIRNN behaves better in high SOC ranges from Figure 3.17(f).

For further illustration, regression analyses and MSE of output error are also calculated, as shown in Figure 3.18. Higher regression coefficients in Figure 3.18(a) mean improved convergence performances relative to the estimated targets, and lower MSE outputs in Figure 3.18(b) mean higher accuracy relative to the estimated targets.

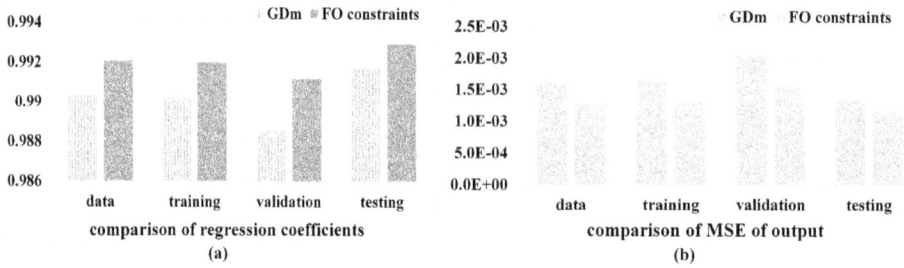

Figure 3.18 Comparison of output performance of fPIRNN with fractional-order constraints to RNN with GDm. (**a**) Regression coefficients and (**b**) MSE of output error. "data" means the entire dataset including training, validation, and testing data. Moreover, "training", "validation", and "testing" are the corresponding results of the three datasets, respectively.

3.3.3.2 Estimation with physics-informed neural network

This section provides the final version of fPIRNN proposed in this chapter, that is, fPIRNN with both FOGDm and fractional-order PDE constraints (marked as "FOGDm and FO loss"). FOGDm is also a proposed method in our work, and PINN only with FOGDm acts as the benchmark for comparisons in this section (marked as "FOGDm"). The convergence process (Figure 3.19a), output portraits (Figure 3.19b,c), and estimation accuracy (Figure 3.19d–f) are provided. With FOGDm and fractional-order constraints, the proposed fPIRNN can converge near $\mathcal{L}_{target}$ much quickly (within 100 epochs), while FOGDm introduces some unstable fractional-order gradients during the convergence process due to the discrete calculation in (3.23). Both "FOGDm" and "FOGDm and FO loss" can bring down the estimated error (within 2.5%), as shown in Figure 3.19f, and fPIRNN with fractional-order constraints still achieves less estimated noise, as shown in Figure 3.19(e).

A regression analysis and MSE of output are provided in Figure 3.20 similar to Figure 3.18. We combine the results of the four NN algorithms together: RNN with GDm ("GDm"), fPIRNN only with fractional-order constraints ("FO constraints"), fPIRNN only with FOGDm ("FOGDm"), and fPIRNN with both FOGDm and fractional-order constraints ("FOGDm and FO loss"). From Figure 3.20, the effectiveness of both FOGDm and fractional-order constraints is shown to achieve higher regression coefficients and a lower MSE than that in Figure 3.18, while fractional-order constraints combined with FOGDm ("FOGDm and FO loss") do not show significant improvements with respect to the performance of the proposed fPIRNN than that only with FOGDm. Considering the results presented from Figure 3.17 to Figure 3.20, some key observations are presented as follows

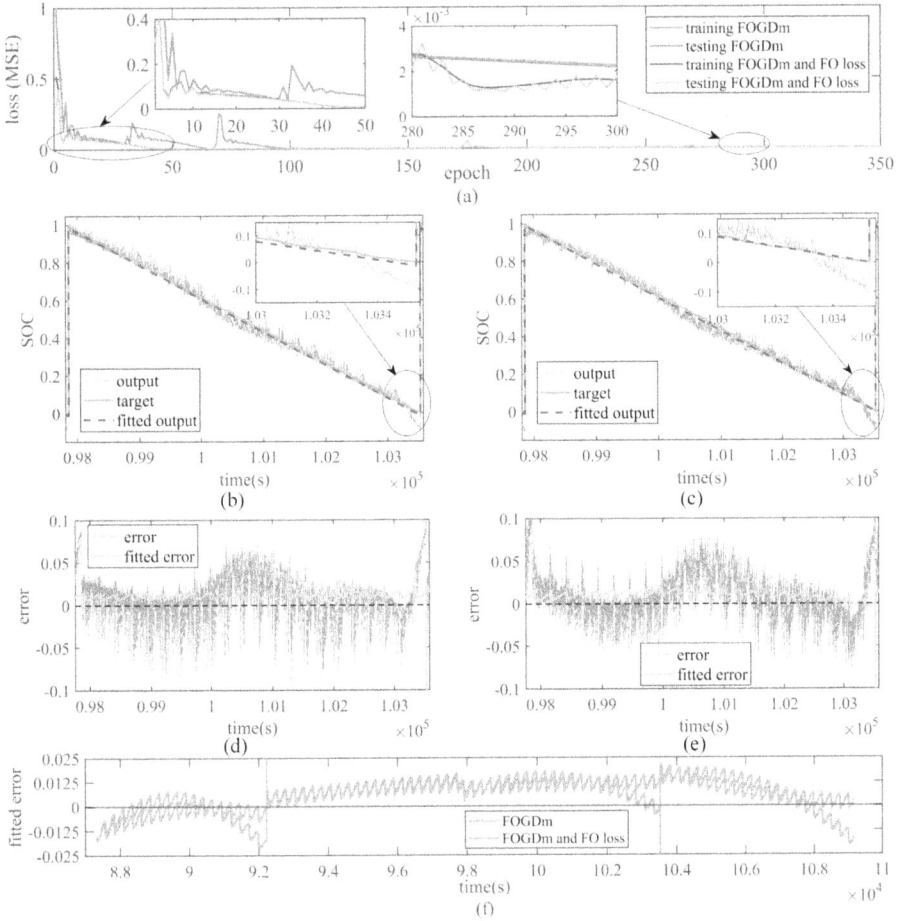

Figure 3.19 Experiment results of fPIRNN with FOGDm and fPIRNN with FOGDm and FO loss. (a) training and testing losses, (b) output SOC with FOGDm, (c) output SOC with FOGDm and FO losses, (d) estimated error with FOGDm, (e) estimated error with FOGDm and FO losses, and (f) comparison of fitted errors.

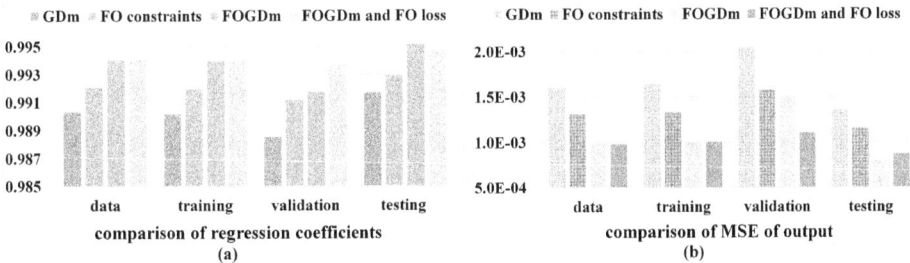

Figure 3.20 Comparison of output performance of four NN algorithms (marked as "GDm", "FO constraints", "FOGDm", and "FOGDm and FO loss"). (a) Regression coefficient and (b) MSE of output.

- The proposed fPIRNN with FOGDm and FO constraints can control SOC's estimation accuracy within 8% with learning noise and within 2.5% with filtering the noise ("fitted error");

- Both FOGDm and fractional-order constraints hold the physics-informed knowledge of LIBs and can optimize the proposed fPIRNN;

- The FOGDm method by backpropagation not only introduces improved performances than fractional-order constraints but also introduces more training fluctuations and estimation noise;

- Besides certain enhancements to fPIRNN, fractional-order constraints can also stabilize the output and reduce the noise in the output;

- FOGDm and fractional-order constraints hold opposite effects on noise, which can compromise with each other, resulting in the final version of fPIRNN in this chapter.

- Combined with the design of embedding physics-informed knowledge, specific conditions at low SOC range and high SOC range can also be embedded to increase the prediction accuracy in these crucial ranges.

3.4 CHAPTER SUMMARY

In this chapter, we presented a physics-informed recurrent neural network for SOC estimations of LIBs by introducing fractional-order gradients and fractional-order constraints into the neural network loss. The fractional-order gradient is designed for weight updates, and the gradient momentum is further added to construct a fractional-order gradient descent method with momentum, named FOGDm. The fractional-order constraint is deduced from the electrochemical relationship described by battery fractional-order models, resulting in a fractional-order PDE form; then, the fractional-order PDE constraint is encoded into the calculation of training loss. Hence, the convergence is improved by embedding battery information, which allows the proposed fPIRNN to learn battery inputs faster and to estimate the outputs more precisely. Moreover, the generalization capability of the ML algorithm is also improved because the battery mechanism acts as a part of the convergence evaluation of the algorithm.

In view of the flexible framework with FOGDm and fractional-order constraints, the proposed fPIRNN can also be applied to other state estimation processes (such as capacity, SOH, and RUL) or safety analysis (such as pre-warning and pack heterogeneity) in addition to SOC estimations. However, the embedded physics in this study is based on a fractional-order ECM, which mainly reflects the electrical characteristics of LIB.

As a potential improvement in the proposed algorithm, PDEs of electrochemical model or other physics-based models with more internal battery mechanisms can be employed to construct unsupervised constraints for the loss function. More inner-state variables and knowledge on their effects related to battery mechanisms could be interesting to consider, not just using the voltage and current signals in this study and this will be the main topic in our future studies.

Bibliography

[1] Ahmed Alsaedi, Bashir Ahmad, and Mokhtar Kirane. A survey of useful inequalities in fractional calculus. *Fractional Calculus and Applied Analysis*, 20(3):574–594, 2017.

[2] Yuquan Chen, Qing Gao, Yiheng Wei, and Yong Wang. Study on fractional order gradient methods. *Applied Mathematics and Computation*, 314:310–321, 2017.

[3] Dongxu Guo, Geng Yang, Xuning Feng, Xuebing Han, Languang Lu, and Minggao Ouyang. Physics-based fractional-order model with simplified solid phase diffusion of lithium-ion battery. *Journal of Energy Storage*, 30:101404, 2020.

[4] Dongxu Guo, Geng Yang, Xuebing Han, Xuning Feng, Languang Lu, and Minggao Ouyang. Parameter identification of fractional-order model with transfer learning for aging lithium-ion batteries. *International Journal of Energy Research*, 45(9):12825–12837, 2021.

[5] Xuebing Han, Xuning Feng, Minggao Ouyang, Languang Lu, Jianqiu Li, Yuejiu Zheng, and Zhe Li. A comparative study of charging voltage curve analysis and state of health estimation of lithium-ion batteries in electric vehicle. *Automotive Innovation*, 2:263–275, 2019.

[6] Taotao Hu, Zheng He, Xiaojun Zhang, and Shouming Zhong. Finite-time stability for fractional-order complex-valued neural networks with time delay. *Applied Mathematics and Computation*, 365:124715, 2020.

[7] Yujiao Huang, Xiaoyan Yuan, Haixia Long, Xinggang Fan, and Tiaoyang Cai. Multistability of fractional-order recurrent neural networks with discontinuous and nonmonotonic activation functions. *IEEE Access*, 7:116430–116437, 2019.

[8] George Em Karniadakis, Ioannis G Kevrekidis, Lu Lu, Paris Perdikaris, Sifan Wang, and Liu Yang. Physics-informed machine learning. *Nature Reviews Physics*, 3(6):422–440, 2021.

[9] Shujaat Khan, Jawwad Ahmad, Imran Naseem, and Muhammad Moinuddin. A novel fractional gradient-based learning algorithm for recurrent neural networks. *Circuits, Systems, and Signal Processing*, 37(2):593–612, 2018.

[10] Shuxian Li, Minghui Hu, Yunxiao Li, and Changchao Gong. Fractional-order modeling and SOC estimation of lithium-ion battery considering capacity loss. *International Journal of Energy Research*, 43(1):417–429, 2019.

[11] Weihan Li, Jiawei Zhang, Florian Ringbeck, Dominik Jöst, Lei Zhang, Zhongbao Wei, and Dirk Uwe Sauer. Physics-informed neural networks for electrode-level

state estimation in lithium-ion batteries. *Journal of Power Sources*, 506:230034, 2021.

[12] Xiaoyu Li, Guodong Fan, Ke Pan, Guo Wei, Chunbo Zhu, Giorgio Rizzoni, and Marcello Canova. A physics-based fractional order model and state of energy estimation for lithium ion batteries. Part I: Model development and observability analysis. *Journal of Power Sources*, 367:187–201, 2017.

[13] J. L. Lovoie, T. Osler, and R. Tremblay. Fractional derivatives and special functions. *SIAM Review*, 18(2):240–268, 1976.

[14] Weilbeer Marc. Efficient numerical methods for fractional differential equations and their analytical background. *Mathematics*, 2005.

[15] Hao Mu, Rui Xiong, Hongfei Zheng, Yuhua Chang, and Zeyu Chen. A novel fractional order model based state-of-charge estimation method for lithium-ion battery. *Applied Energy*, 207:384–393, 2017.

[16] Achraf Nasser-Eddine, Benoît Huard, Jean-Denis Gabano, and Thierry Poinot. A two steps method for electrochemical impedance modeling using fractional order system in time and frequency domains. *Control Engineering Practice*, 86:96–104, 2019.

[17] Wei Peng, Jun Zhang, Weien Zhou, Xiaoyu Zhao, Wen Yao, and Xiaoqian Chen. IDRLnet: A physics-informed neural network library. *arXiv preprint arXiv:2107.04320*, 2021.

[18] Dian Sheng, Yiheng Wei, Yuquan Chen, and Yong Wang. Convolutional neural networks with fractional order gradient method. *Neurocomputing*, 408:42–50, 2020.

[19] Jinpeng Tian, Rui Xiong, Jiahuan Lu, Cheng Chen, and Weixiang Shen. Battery state-of-charge estimation amid dynamic usage with physics-informed deep learning. *Energy Storage Materials*, 50:718–729, 2022.

[20] Jinpeng Tian, Rui Xiong, and Quanqing Yu. Fractional-order model-based incremental capacity analysis for degradation state recognition of lithium-ion batteries. *IEEE Transactions on Industrial Electronics*, 66(2):1576–1584, 2018.

[21] Qi Zhang, Yunlong Shang, Yan Li, Naxin Cui, Bin Duan, and Chenghui Zhang. A novel fractional variable-order equivalent circuit model and parameter identification of electric vehicle Li-ion batteries. *ISA Transactions*, 97:448–457, 2020.

[22] Changfu Zou, Xiaosong Hu, Satadru Dey, Lei Zhang, and Xiaolin Tang. Nonlinear fractional-order estimator with guaranteed robustness and stability for lithium-ion batteries. *IEEE Transactions on Industrial Electronics*, 65(7):5951–5961, 2017.

CHAPTER 4

Fractional-Order Algorithms for Battery Capacity Estimation

To handle and manage battery degradation in electric vehicles (EVs), various capacity estimation methods have been proposed and can mainly be divided into traditional modeling methods and data-driven methods. For realistic conditions, data-driven methods take the advantage of simplicity in implementation. However, state-of-the-art machine learning (ML) algorithms are still kinds of black-box models; thus, the algorithms do not have a strong ability to describe the inner reactions or degradation information of batteries. Due to the lack of interpretability, ML may not learn the degradation principle correctly and may need to rely on big data quality.

In this chapter, we apply fractional calculus to a physics–informed recurrent neural network (PIRNN) for fast battery degradation estimation of running EVs. The proposed fractional-order neural network is designed to learn battery degradation mechanisms. Incremental capacity analysis (ICA) is firstly conducted to extract aging characteristics, which is selected as the input of the algorithm. The fractional–order gradient descent (FOGD) method is also applied to improve the training convergence and embedding of battery information during backpropagation; then, the recurrent neural network is selected as the main architecture of the algorithm. A battery dataset of ten EVs with a total 5697 charging snippets is constructed to validate the performance of the proposed algorithm. Experimental results show that the proposed PIRNN with ICA and the FOGD method could achieve the relative error within 5% for most snippets of the ten EVs. The algorithm could even achieve a stable estimation accuracy (relative error < 3%) during the first three-quarters of a battery's lifetime.

4.1 BACKGROUNDS

As the most commonly used energy supply for electric vehicles (EVs), the durability of lithium–ion batteries (LIBs) is a key concern for the intelligent management of

DOI: 10.1201/9781003670902-4

batteries [29, 46]. Batteries degrade according to dynamic operation conditions and environmental conditions during their lifetime; thus, capacity estimation methods for LIBs have been widely designed recently [20, 35]. Currently, researchers have proposed various kinds of model-based methods and data-driven methods [15, 23, 44] for the estimation of the state of charge (SOC) [30], state of health (SOH) [34, 14], and remaining useful life (RUL) [15, 23, 44, 45]. Data–driven methods include statistical methods such as entropy, correlation, and wavelet transform [6, 22] and machine learning (ML) methods such as regression models, neural networks (NNs), reinforcement learning, and even transfer learning [5, 50, 51]. ML is gradually applied to battery capacity estimation and degradation prediction in an intelligent way to make an algorithm learn the battery degradation trend [37]. Among ML algorithms, NNs are a widely investigated and easily employed class for estimating the battery capacity of EVs [28], such as recurrent neural networks (RNNs), long short-term memory (LSTM) [14], and gated recurrent units (GRUs) [7]. Recently, researchers have proposed several innovative ML methods for EVs and hybrid electric vehicles (HEVs), such as optimal energy management strategies [18], supervised ML for battery health performance [8], deep learning for seasonal EV forecasting models [9], and so on.

For the convergence of the loss function and weight updates of NNs, the first-order optimization method (gradient descent based) and second-order method (Hessian matrix based) are developed during backward process of network training. The calculation of the Hessian matrix may use heavy computing resources and storage; thus, its application is limited or replaced by the quasi-Newton method [1, 33]. The gradient descent (GD)-based method is currently the main optimization method, and the basic GD method is improved in several well-known GD methods, such as stochastic gradient descent (SGD), mini batch gradient descent, GD with momentum, the Nesterov accelerated gradient (NAG), adaptive gradient (Adagrad), adaptive moment estimation (Adam), and so on [4, 2]. Moreover, researchers have proposed more innovative gradient-based methods, such as smoothed functional algorithms with a norm-limited update vector [26], grafting gradient descent [49], normalized simultaneous perturbation stochastic approximation (SPSA) [3], the Gaussian–Stein variational gradient descent [24], and so on. While research on GD is thriving and includes a large number of methods, this chapter focuses on the fractional–order information of batteries combined with GD methods; thus, the basic GD method is first selected in this study.

With careful preprocessing and excellent calculation support, ML algorithms can achieve perfect performance in capacity estimation accuracy at the lab level. However, in realistic situations, these kinds of algorithms cannot exhibit their advantages due to limited data quality [10]. With unstable and dynamic operation conditions, battery degradation may vary from the aging situation in the lab and cannot be predicted by pre-trained algorithms at the lab level [28]. Hence, it is necessary to make ML learn battery mechanisms. In this way, an algorithm can understand the inner reactions during battery degradation from the aspect of first principles. In the literature, physics–informed machine learning (PIML) has been proposed to solve problems related to objects with physical laws [19]. PIML can encode the physical law of a predicted object into observational bias, inductive bias, and learning bias [19, 21, 36]; thus, it combines the advantages of model-based models and data-driven algorithms [41]. Informed by battery knowledge from batteries' physical laws, the neural network can be enhanced from different aspects, such as the state's feedback, the loss function, and the gradient after backpropagation, for application to LIBs [42, 43, 39]. Hence, battery information needs to find a way to be embedded into learning algorithms, and fractional calculus needs to come out of the fractional dynamics in the battery. Fractional calculus (FC) is first applied to LIBs to enhance non-linear fractional–order (FO) modeling [25, 38, 48, 52]. The reflection of the diffusion dynamics of battery modeling is described by introducing a fractional-order element called the Warburg element [27, 40]. Then, fractional-order methods, such as the FO extended Kalman filter (EKF) [32] and FO unscented Kalman filter (UKF) [47], are proposed for the estimation of battery states. Moreover, to process time–series data and their characteristics, the fractional-order recurrent neural network (FORNN), fractional–order state feedback, and the fractional-order gradient have also been investigated theoretically [16, 17, 43]. Since the fractional-order gradient can introduce history gradient data to improve the local optimum problem, the basic GD method is extended to the fractional-order gradient descent (FOGD) method, which can be applied to RNNs to characterize the fractional-order dynamical battery degradation information in time series.

In this chapter, a physics–informed method is proposed to enhance algorithm interpretability. Incremental capacity analysis (ICA) is also conducted to extract degradation characteristics as the inputs for machine learning. Then, since the fractional-order element (constant phase element, CPE) can reflect the charge transfer reaction in the mid-frequency and solid diffusion dynamics in the low frequency of LIBs [27, 40], a fractional-order gradient descent (FOGD) method is applied to improve the convergence of the algorithm during training. To process the time–series battery data, the main architecture of the algorithm is chosen to be an RNN. It should be noted that the main algorithm proposed in this chapter is not limited to an RNN, and other suitable ML algorithms can be applied, such as LSTM, GRUs, or even meta-learning. Similarly, the FOGD method proposed in this chapter can be extended to other kinds of GD methods, such as SGD, GD with momentum, and so on.

Here are several main contributions of the proposed fractional–order algorithm.

- A physics–informed recurrent neural network (PIRNN) with ICA and FOGD methods is proposed for capacity estimation of LIBs. The peak magnitudes of the IC curves are extracted as input characteristics for the neural network, and a fractional–order gradient is applied.

- The proposed PIRNN can learn the battery information of running EVs during training convergence and achieve stable capacity estimation accuracy (relative error $< 3\%$) using a $LiFePO_4$ battery dataset over the first three-quarters of a lifetime. This demonstrates that the proposed PIRNN can be applied to realistic batteries from running EVs and hold its accuracy.

- A battery dataset is constructed based on ten running EVs covering the years 2018–2022 and with over a 40,000 km mileage. The dataset contains 5697 charging snippets and covers almost the whole battery lifetime, which can be the validation dataset for the machine learning algorithms.

4.2 FRACTIONAL-ORDER INFORMATION OF BATTERY

The basic knowledge of fractional calculus has been introduced in Chapters 2 and 3, such as fractional–order derivative in Caputo definition and fractional–order gradient. This chapter focuses on the employment of fractional-order battery information into RNN, resulting in a fractional-order algorithm named physics–informed recurrent neural network (PIRNN), for capacity estimation of battery. Hence, we directly present the fractional-order gradient with Caputo definition in the following

$$x_{k+1} = x_k - \rho \nabla {}^{C}_{x_{k-1}} D^{\alpha}_{x_k} f(x), \tag{4.1}$$

where k is the iteration step and x_k is the discrete value in the step k, ${}^{C}D^{\alpha}_{x}$ is the fractional operator in the Caputo definition, $0 < \alpha < 1$, and $f(x)$ is a smooth convex function with its integer-order gradient, usually presented as $\nabla f(x)$. Suppose that (4.1) can converge to the global extreme point x^*, the convergence of the fractional-order gradient in (4.1) depends on the fractional order α, the learning rate ρ and initial value x_0. Fractional-order gradients are applied to the gradient descent method for the neural network. Before introducing the algorithm, the fractional–order characteristics of LIBs are presented in this section.

4.2.1 EV dataset and battery degradation

For battery durability, state estimations are quite important, such as the state of charge (SOC), which can be obtained by a look-up table with open circuit voltage (OCV), and state of health (SOH), which illustrates the real-time battery charging/discharging ability. The battery capacity is the basis of the SOH. For realistic data, battery capacity can be calculated as

$$Q_{\alpha,\beta} = \frac{\Delta Ah}{\Delta SOC} = \frac{\int_{t_\alpha}^{t_\beta} I(\tau)d\tau}{SOC(t_\beta) - SOC(t_\alpha)} \tag{4.2}$$

where t_α and t_β are two separate moments and $I(t)$ is the current during (t_α, t_β). Equation (4.2) is based on the Ampere-hour integral method and depth of charge (DOC). The DOC is the SOC change range during charging. The labels used in this chapter are deduced and based on equation (4.2) with the calibration of the SOC–OCV curve.

In general, Ni-CO-Mn (NCM) batteries and LiFePO$_4$ (LFP) batteries are the two major kinds of LIBs used. The available battery in this chapter is an LFP battery. A typical degradation fitting curve of an LFP battery under regular cycles is usually a power-law function, as shown in Figure 4.1. It shows a slowed-down degradation when the battery goes into the late period of its lifetime. However, under the dynamic operation conditions in EVs, the data obtained in this chapter are for a set of batteries with faster degradation rather than a slow power function.

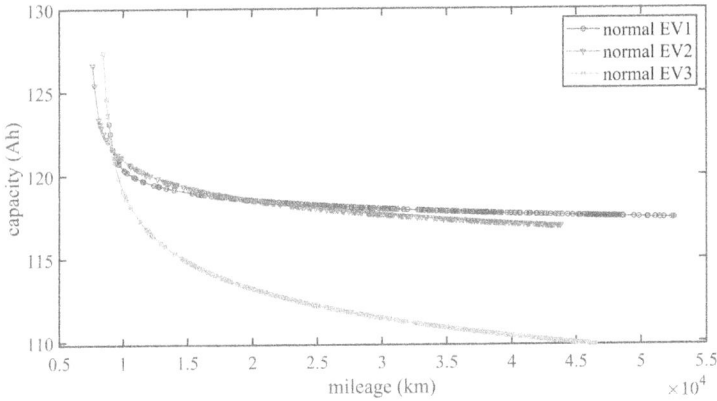

Figure 4.1 Typical degradation curves of LiFePO$_4$ (LFP) batteries with mileage.

The dataset used in this chapter comes from our co-operation project, which contains 10 EVs named from "EV1" to "EV10" due to VIN privacy protection. Ten EVs numbered from 1 to 10 are sampled with the operation data of their power battery; then, the battery capacity is calculated, based on equation (4.2) with the calibration of the SOC–OCV curve. The basic information of the ten EVs (EV1–EV10) is listed in Table 4.1. The battery cell used in the ten EVs is a kind of LFP cell with a 130 Ah rated capacity and 14.9 kWh rated power, and the rated voltage range is from 2.5 V to 3.65 V. The battery pack of the EV contains 36 cells connected in series. Due to the privacy protection, we could only obtain the basic rated parameter and OCV-SOC map of this LFP cell provided by our co-operation company. The driving cycles of the ten EVs are mostly urban operating conditions with various drivers instead of experimentally designed conditions. Since dynamic operation conditions result in fluctuating vehicle data, charging snippets from the EVs are selected and data pre-processing is conducted to filter stable data to construct the battery dataset used in this chapter. From Table 4.1, we see that the time range of the filtered battery dataset covers from 2018 to 2022 and the vehicle mileage varies from 87 to 52,992 km, which contain a long enough period and enough mileage for representing an aging LFP battery. The charging snippets are filtered by the SOC upper limit (>80%) and lower limit (<40%) so that the depth of charge (DOC) could be larger than 40%; then, the battery degradation characteristics could be deduced from the voltage curves. Although a tough filter condition is applied, the obtained snippets are enough to construct a battery dataset for degradation study.

We could not obtain all the accurate capacity labels from standard capacity test, since these realistic running vehicles are possessed by the customers. Hence, based on equation (4.2), Figure 4.2 shows the calculated capacities used as labels in this chapter, and the rated capacity of this battery pack in these ten EVs is 130 Ah. Compared to Figure 4.1, the battery degradation shown in Figure 4.2 presents faster trends during the battery lifetime. Based on the obtained data, this may be caused

Table 4.1 Battery dataset extracted from charging snippets of ten vehicles

No.	Snippets Amount	Time Range	Mileage Range (km)	SOH Range (Ah)
EV1	698	2018.09.18-2022.02.28	8730-52992	124.9076-97.9608
EV2	290	2018.10.11-2022.03.29	87-21987	131.0177-115.6695
EV3	660	2018.09.22-2022.02.28	4400-47461	128.4828-102.3552
EV7	571	2018.12.10-2022.01.31	7605-43792	123.3471-105.1326
EV5	633	2018.09.19-2021.09.16	8355-46569	120.4993-105.9154
EV6	562	2018.09.18-2022.03.02	8395-45933	122.0545-107.6695
EV7	650	2018.09.18-2022.03.01	7725-46737	115.7236-108.9408
EV8	487	2018.09.18-2022.01.31	8165-40186	123.4687-88.0738
EV9	590	2018.09.18-2021.07.20	7660-45148	121.3203-107.3435
EV10	556	2018.09.18-2021.12.20	8538-44569	119.3115-110.6705

by the chemical composition of this type of LFP battery or caused by the dynamic operation conditions. The fast degradation mentioned in this chapter produces the aging curves shown in Figure 4.2, which are different from the general power-law trend. It should be noted that we do not discuss the specific capacity plunge (dramatic capacity decrease) in this chapter, although it can be observed that EV8 in Figure 4.2 holds an abnormal aging curve with an almost 30% degradation. The battery capacity plunge is another worthy investigation topic in our other works, and our work in this chapter is the employment of the PIRNN with a physics-informed method with this dataset containing ten EVs.

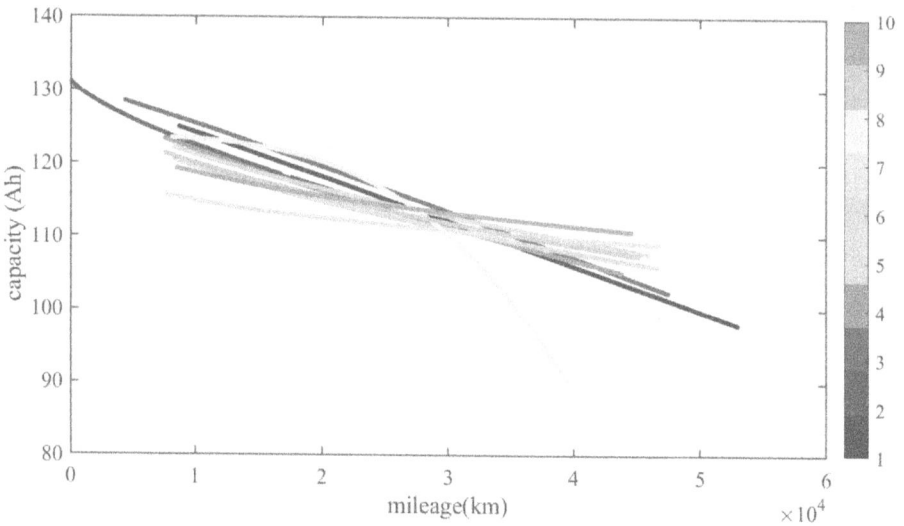

Figure 4.2 The battery capacity labels in battery dataset of the ten EVs.

4.2.2 Fractional-order equivalent circuit model

An electrochemical model with partial differential equations (PDEs) can describe the physical laws of LIBs at the microscopic particle level, such as lithium ion concentrations, potential distributions, and interface Bulter–Volmer kinetics; however, the particle-level reactions are coupled with each other and can hardly be used by NNs [13]. Compared to electrochemical models, equivalent circuit models (ECMs) can describe the main electrical reactions with decoupled equations and simpler structures, which contain enough physical battery information for NNs. The fractional–order constant phase element (CPE), also called the Warburg element [11, 12], is introduced due to the fractional-order capacitance of LIBs in low frequencies [48]. The voltage–current relationship in the time domain and impedance in the frequency domain of the CPE are presented as

$$
\begin{cases}
i(t) = C_{CPE}\frac{\mathrm{d}^\alpha u(t)}{\mathrm{d}t^\alpha}, \ 0 < \alpha < 1, t \geq 0 \\
Z_{CPE}(s) = \frac{U(s)}{I(s)} = \frac{1}{C_{CPE} \cdot s^\alpha} = \frac{1}{C_{CPE}(j\omega)^\alpha}
\end{cases}, \tag{4.3}
$$

where Z_{CPE} is the complex impedance, C_{CPE} is the capacity coefficient, j is the imaginary unit, α is the fractional order related to capacitance dispersion, and ω is the angular frequency.

The battery EIS mainly contains three parts: a high-frequency inductive tail, a mid-frequency reaction, and low-frequency diffusion dynamics. By employing the CPE to describe the fractional-order characteristics of the solid diffusion dynamics in the low-frequency range [27], a fractional-order model (FOM) can be deduced. The fractional–order Partnership for a New Generation of Vehicles (PNGV) model is a typical second-order FOM [27], as shown in Figure 4.3, and it is widely used due to its full-scale reflection of LIB dynamics in all frequencies.

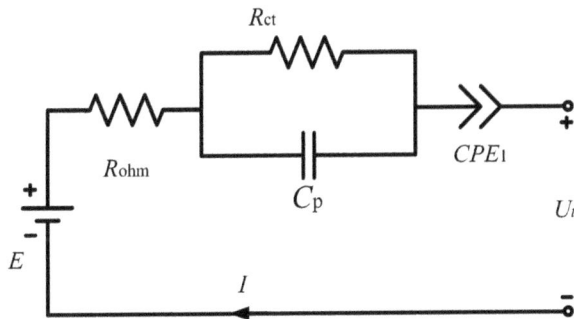

Figure 4.3 The fractional-order PNGV model of LIBs for physical battery information embedded in the PIRNN.

The state-space model (SSM) of the fractional-order PNGV model is [27]

$$\begin{bmatrix} \frac{dU_{C_p}}{dt} \\ \frac{d^{\alpha_1} U_{CPE_1}}{dt^{\alpha_1}} \end{bmatrix} = \begin{bmatrix} -\frac{1}{R_1 C_p} & 0 \\ 0 & 0 \end{bmatrix} \begin{bmatrix} U_{C_p} \\ U_{CPE_1} \end{bmatrix} + \begin{bmatrix} \frac{1}{C_p} \\ \frac{1}{C_{CPE_1}} \end{bmatrix} I,$$

$$U_t = \begin{bmatrix} 1 & 1 \end{bmatrix} \begin{bmatrix} U_{C_p} \\ U_{CPE_1} \end{bmatrix} + R_{ohm} I + E, \tag{4.4}$$

where U_{CPE_1} and U_{C_p} are the voltages of the Warburg element CPE_1 and C_p; C_{CPE_1} and C_p are the capacitance of CPE_1 and C_p; α_1 is the fractional order of CPE_1; E is the open circuit voltage (OCV); U_t is the terminal voltage; and I is the current. Based on the SSM of the fractional-order PNGV model in Equation (4.4), the impedance of LIBs can be presented as

$$Z = R_{ohm} + Z_{ARC} + Z_{Warburg} = R_{ohm} + \frac{R_{ct}}{R_{ct} C_p s + 1} + \frac{1}{C_1 s^{\alpha_1}} \tag{4.5}$$

where R_{ohm} is the Ohmic resistance in high-frequency ranges, the RC tank Z_{ARC} (R_{ct} and C_p) is the double-layer effects and charge-transfer reactions in the mid-frequency range, and $Z_{Warburg}$ (CPE_1) is the solid diffusion dynamics (fractional-order characteristic) in the low-frequency range, respectively.

4.2.3 Incremental capacity analysis

To solve the capacity estimation of EVs with the degradation shown in Figure 4.2, only commonly used variables, such as current, voltage, and the SOC, are not enough. Inner knowledge of battery degradation should be further investigated. Incremental capacity analysis is always applied in the electrochemical modeling and capacity loss of LIBs. ICA is the variation trends of charge quantity in the cell voltage, which means the cell's ability to store charge quantity in every voltage period. Since the working voltage range of a cell is certain, the ability of quantity storage illustrates the available capacity.

Before the calculation of ICA, a cell inside the EV pack should have been selected to fetch the corresponding cell voltage to extract the ICA. However, the EV battery pack contains 36 cells. Hence, we statistically analyzed the maximum–voltage cell and the minimum–voltage cell marked by the battery management system (BMS) of the ten vehicles, and part of the maximum and minimum cell results of EV1 and EV7 is shown in Figure 4.4. To ensure the integration of the voltage plateau in the LFP cell and the corresponding peaks in the IC curves, the cell with the maximum voltage in every EV is selected because this cell may have been the first one to reach the charging cut-off voltage.

During its whole lifetime, EV7 contains 650 charging snippets totally in different DOCs, as shown in Figure 4.5, which contains enough degradation information for algorithm learning. To present the ICA relationship with battery degradation clearly, 55 snippets ranked from 1 to 55 according to the cycling increase are extracted from the 650 snippets by every 13 snippets with data preprocessing. Figure 4.6(a,b) exhibit the charging voltage curves extracted from the selected LFP cell of EV7

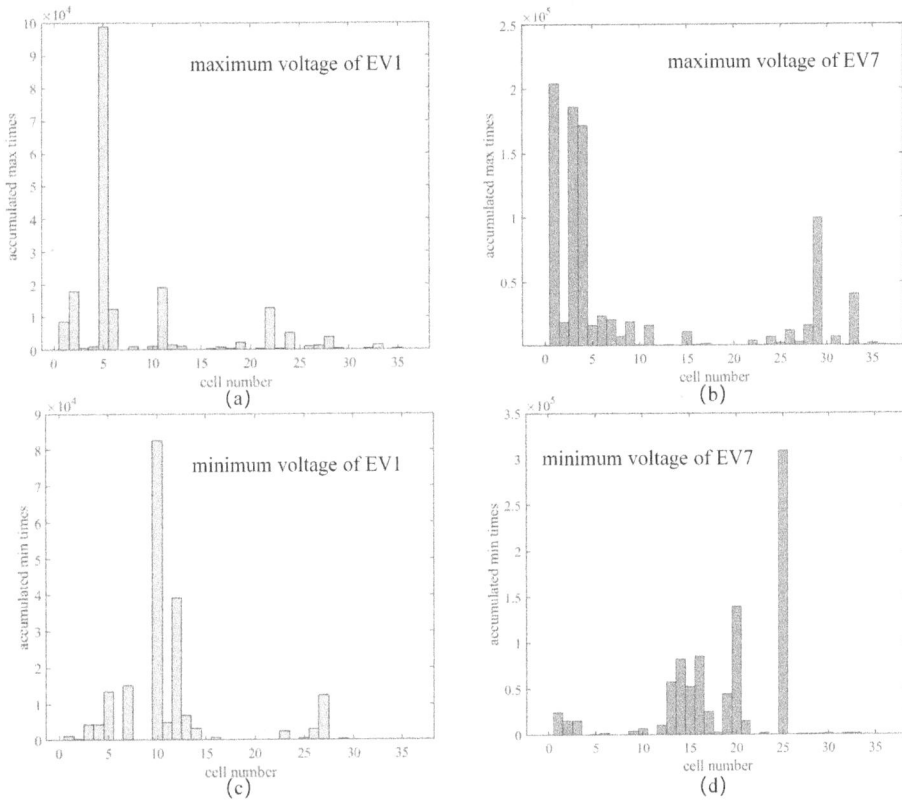

Figure 4.4 The accumulated time distribution of the maximum voltage cell marked by the battery management system (BMS) during all charging snippets of EV1 and EV7.

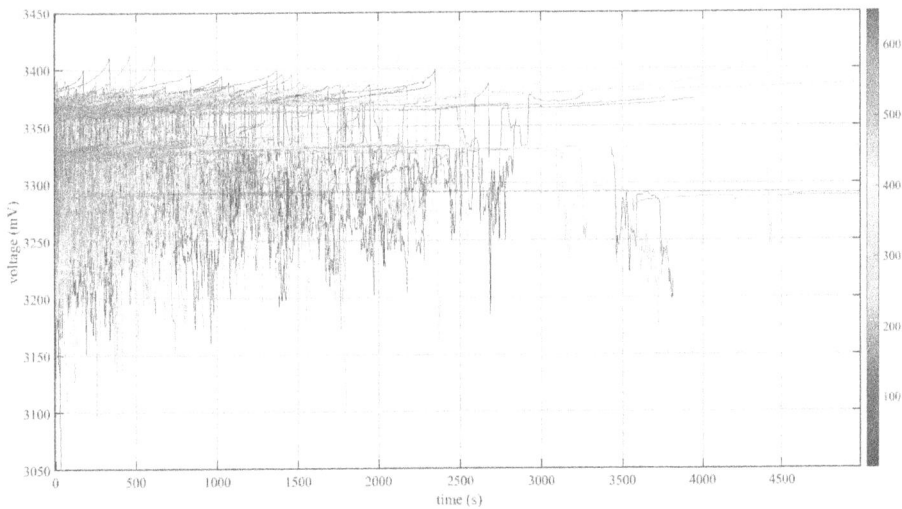

Figure 4.5 The 650 charging voltage curves of the selected LFP cell (the cell with maximum voltage) in EV7 with degradation.

and the corresponding IC curves, respectively. To highlight the ICA effects, only the two plateau regions of the LFP cell are plotted in Figure 4.6(a), and the peaks in the IC curves shown in Figure 4.6(b) can be clearer. Figure 4.6(a) presents the shift in the voltage curves to the left region of the figure, and the lengths of the two plateau regions decrease with a cycling increase. Hence, the lengths of the two plateau regions are related to the battery degradation, and they can act as an indicator for RNN training. Then, the voltage plateau is transferred to the IC curves, as shown in Figure 4.6(b), which demonstrates that the peaks' magnitude in the IC curves decreases when the battery cycle increases. Hence, this work chooses the two peaks' magnitude as the inputs of the proposed algorithm.

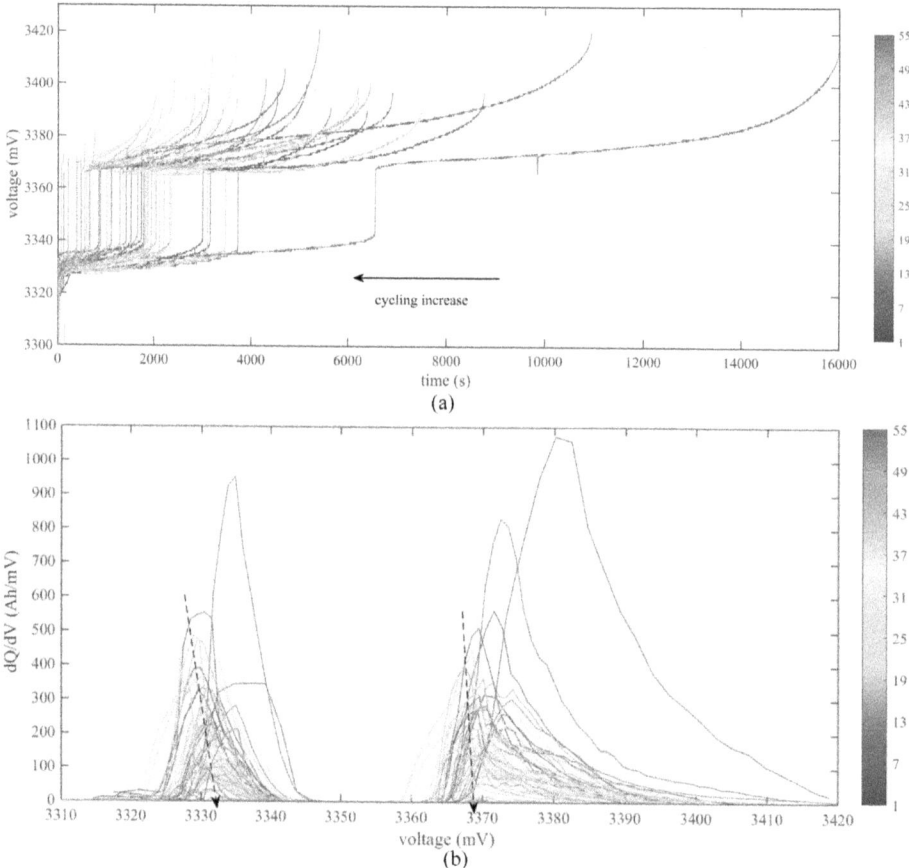

Figure 4.6 The extracted voltage curves and incremental capacity (IC) curves of the selected LFP cell in EV7 during charging with degradation. (**a**) The extracted charging voltage curves with two extracted plateau regions; (**b**) the IC curves.

4.3 FRACTIONAL-ORDER PHYSICS-INFORMED RECURRENT NEURAL NETWORK

4.3.1 Physics-informed input with ICA characteristics

The flowchart and framework for the whole structure of the PIRNN is shown in Figure 4.7. For battery degradation estimation, ICA with degradation information and a fractional-order gradient are both applied to the neural network, resulting in a physics-informed algorithm. The ICA inputs, network structure, and fractional-order gradient descent(FOGD) method will be introduced in the following sections, respectively.

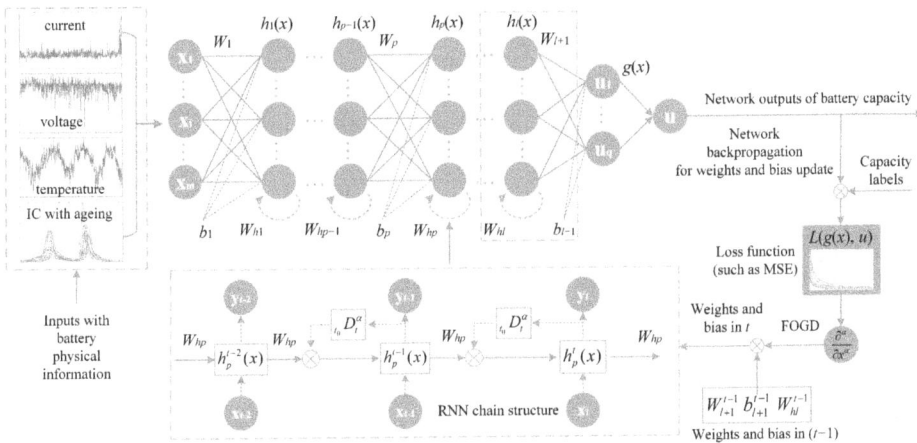

Figure 4.7 The flowchart and framework of the proposed PIRNN for fast battery degradation modeling.

The ICA curves of the ten EVs are extracted from the battery voltage curves, and part of the ICA curves are shown in Figure 4.8. For the LFP battery, the position and the length of the two plateau regions in the voltage curve shift when the battery cycles increase. From the battery mechanism aspect, the two plateau regions in the voltage curve correspond to the two peaks in the IC curves, which demonstrates the change in the loss of lithium inventory (LLI) or loss of active material (LAM) with battery aging. Hence, the length of the two plateau regions related to the battery degradation can be transferred to the magnitude of the two peaks, which are called the peak1 magnitude and peak2 magnitude in the following section, as marked in Figure 4.8. Although different cells may have different IC curves, as shown in Figure 4.8, the trends related to battery degradation are similar, and thus a normalization process is applied for every cell before they are input into the network.

4.3.2 Physics-informed structure of RNN

Besides the ICA characteristic, a physics-informed algorithm could be achieved from the structure aspect. The fractional-order RNN in Figure 4.9 contains an input layer with m neurons, the hidden layers $(h_1, \ldots, h_{p-1}, h_p, \ldots, h_l)$ with $n_p(p = 1, 2, \ldots, l)$

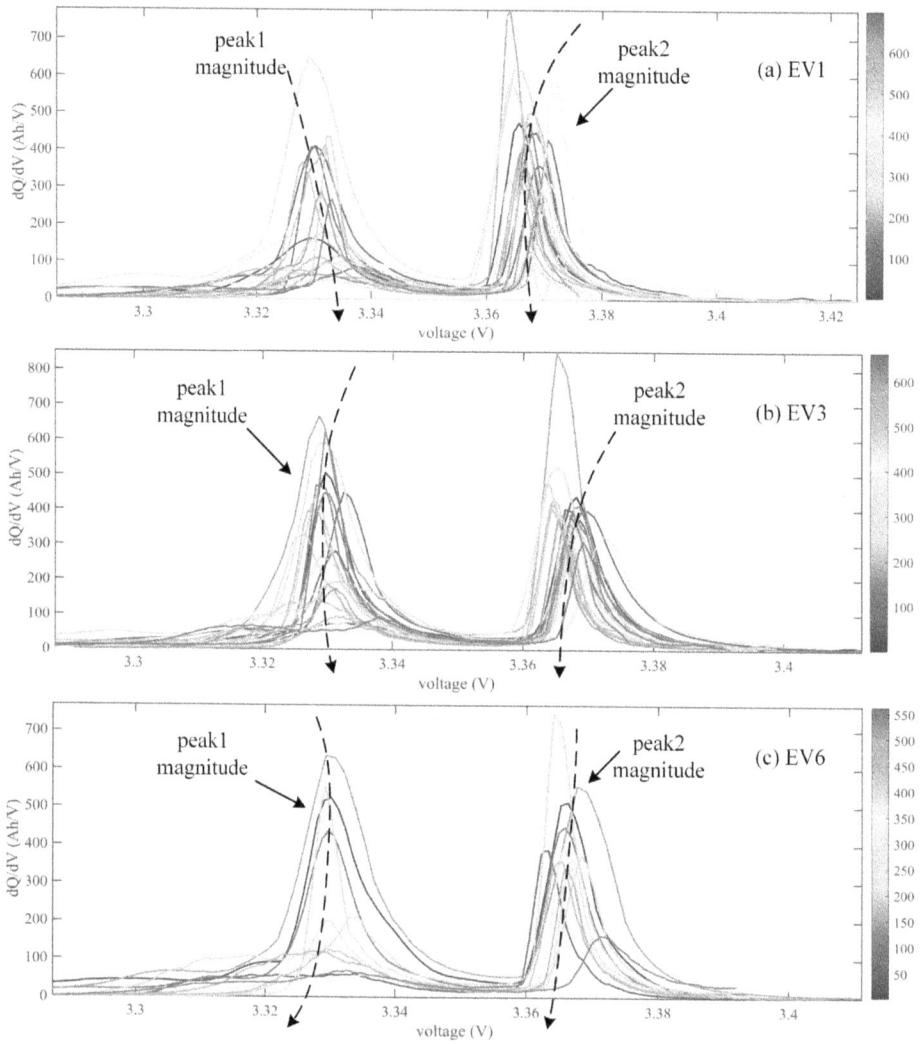

Figure 4.8 The IC curves and characteristics extracted from the IC curves of EV1, EV3, and EV6.

neurons, and an output layer with q neurons, respectively. Suppose that the training data are $(\mathbf{x}_i, \mathbf{u}_i), i = 1, 2, \ldots, N$, where $\mathbf{x}_i = (x_{i1}, x_{i2}, \ldots, x_{im})^{\mathrm{T}}$ is the input and $\mathbf{u}_i = (u_{i1}, u_{i2}, \ldots, u_{iq})^{\mathrm{T}}$ is the ideal output. The vectors $\mathbf{x}_i$ and $\mathbf{u}_i$ are presented as x and u in the following context. In the chain structure of the RNN, W_p and b_p are the weight and bias matrix connecting the $(p-1)$th hidden layer to the pth hidden layer, respectively W_{hp} is the weight for the memory updates of the pth hidden layer, $g(x)$ is the non-linear activation function, and $L(g(x), u)$ is the loss function. Within the pre-set maximum epochs, the RNN went through forward propagation and backpropagation with the training dataset. The forward propagation starting

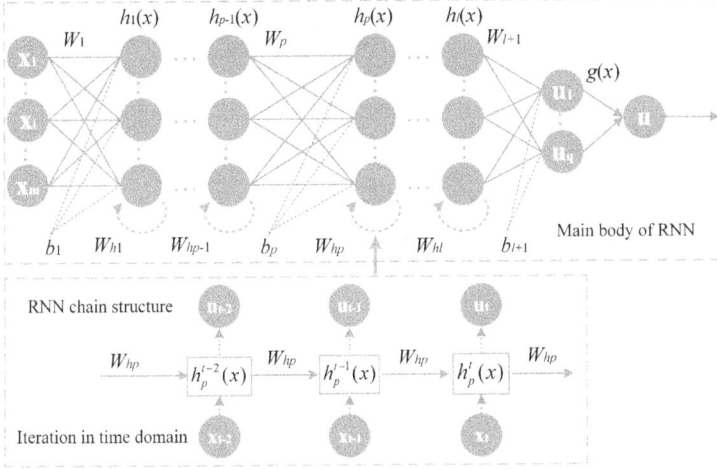

Figure 4.9 The structure of the proposed fractional-order recurrent neural network.

from the input layer can be presented as

$$
\begin{cases}
a_p(x) = W_p h_{p-1}(x) + b_p, \\
h_p(x) = g(a_p(x)), p = 1, 2, \ldots, l
\end{cases}
\tag{4.6}
$$

where $a_p(x)$ and $h_p(x)$ are the input and the output of the pth hidden layer, respectively. The weights W_p and W_{hp} and bias b_p can be updated from the output layer to the hidden layers, and it always makes the weights $W_p = W_{hp}$ in real applications. The loss function $L(g(x), u)$ applied in this chapter (also called the performance function) is selected as the mean squared error (MSE), which is calculated as

$$
L(g(x), u) = \frac{1}{N_{\text{data}}} \sum_{i=1}^{N_{\text{data}}} (u_i(x_i, t_i) - \hat{u}_i)^2
\tag{4.7}
$$

where N_{data} is the battery data amount and $\hat{u}_i$ is the actual labels of the battery capacity.

4.3.3 Fractional-order gradient descent method

The fractional-order gradient descent (FOGD) method is proposed in Chapter 3 and our published article [43]. Here, we directly present the results of the fractional-order gradient method in this section, and detailed deduction processes can be found in Chapter 3. The FOGD for the updates of the weight W_p is presented as

$$
\begin{aligned}
W_p^{k+1} &= W_p^k - \eta \cdot {}_{W_p^0} D_{W_p^k}^\alpha L(g(x), y) \\
&= W_p^k - \eta \cdot \frac{\partial^\alpha L(g(x), y)}{\partial \left(W_p^k\right)^\alpha}
\end{aligned}
\tag{4.8}
$$

where $\partial^\alpha L(g(x), y)/\partial(W_p^k)^\alpha$ is the fractional-order gradient of the weight W_p of the loss function $L(g(x), y)$ and η is the learning rate (iteration step size). Based on

$a_p(x) = W_p h_{p-1}(x) + b_p$ in equation (4.6) and the chain rule [31], equation (4.8) is deduced as

$$W_p^{k+1} \approx W_p^k - \eta \cdot \frac{\partial L(g(x), y)}{\partial a_p^k(x)} \cdot \frac{\partial^\alpha a_p^k(x)}{\partial \left(W_p^k\right)^\alpha} \tag{4.9}$$

Remark 4.1: It should be noted that the chain rule used in (4.9) is an approximation result [31]. Using the generalized Leibniz series expansion, the fractional derivative of a composite function $f(g(x))$ should have an infinite term as

$$D^\alpha f(g(x)) = \sum_{k=0}^{\infty} \frac{\Gamma(1+\alpha)}{\Gamma(1+k)\Gamma(1+\alpha-k)} g^{(\alpha-k)}(x) f^{(k)}(g(x)) \tag{4.10}$$

where $g^{(\alpha-k)}(x)$ represents the fractional derivatives of the inner function $g(x)$. The chain rule for the fractional derivative of a composite function $f(g(x))$ can only obtain numerical approximation within a certain range of x and would have large errors when x increases. Hence, the use of chain rule should be carefully considered.

In equation (4.9), the gradient of the input $a_p^k(x)$ to the loss function $(\partial L(g(x), y)/\partial a_p^k(x))$ can be obtained by

$$\begin{cases} \dfrac{\partial L(g(x), y)}{\partial h_p^k(x)} = W_{p+1}^k \dfrac{\partial L(g(x), y)}{\partial a_{p+1}^k(x)} \\[3mm] \dfrac{\partial L(g(x), y)}{\partial a_p^k(x)} = \dfrac{\partial L(g(x), y)}{\partial h_p^k(x)} \cdot g'(a_p^k(x)) \\[3mm] \qquad\qquad = W_{p+1}^k \dfrac{\partial L(g(x), y)}{\partial a_{p+1}^k(x)} \cdot g'(a_p^k(x)) \end{cases} \tag{4.11}$$

where $\partial L(g(x), y)/\partial h_p^k(x)$ and $\partial L(g(x), y)/\partial a_p^k(x)$ are the gradients of the output $h_p^k(x)$ and the input $a_p^k(x)$ of the pth hidden layer, respectively. Combining equation (4.11) and the Caputo derivative, the fractional-order gradients of the weight W_p of the loss function $L(g(x), y)$ can be deduced as

$$\frac{\partial^\alpha L(g(x), y)}{\partial (W_p^k)^\alpha} = \frac{h_{p-1}^k(x)}{\Gamma(2-\alpha)} \cdot \frac{\partial L(g(x), y)}{\partial a_p^k(x)} \left(W_p^k - W_p^0\right)^{1-\alpha} \tag{4.12}$$

where W_p^0 is the initial value of the weight W_p. Taking equation (4.12) into equation (4.9), we can obtain the FOGD method as

$$W_p^{k+1} = W_p^k - \eta \cdot \frac{h_{p-1}^k(x)}{\Gamma(2-\alpha)} \cdot \frac{\partial L(g(x), y)}{\partial a_p^k(x)} \left(W_p^k - W_p^0\right)^{1-\alpha} \tag{4.13}$$

where α is a fractional order that may be related to battery knowledge and sensitive to the training results. Equation (4.13) is the basic equation of the FOGD method to update weights in the backpropagation process.

With the updated equation (4.13), the procedure of the PIRNN with the maximum epoch (E_{max}) can be summarized as follows.

- Step 1: Perform initialization, with $W_p^0, p = 1, 2, \ldots, l, b_p^0$, the learning rate l_r, the fractional order α, and so on.

- Step 2: Obtain the battery dataset, and then preprocess the data (including extraction of inputs such as ICA magnitude) and divide the whole battery dataset into training, validating, and testing datasets.

- Step 3: Perform the feedforward computation process, in a discrete form, with data flows in the PIRNN from the input layer to output layer, and then calculate the MSE (mse^k).

- Step 4: Perform the backpropagation computation process, starting from the last output layer, to calculate the gradients between layers, and then update the weight W_p and bias b_p with (4.13).

- Step 5: Perform validation. Check if mse^k satisfies the target value or if the maximum epoch E_{max} has arrived. If so, go to Step 6; if not, go to Step 3 and $epoch = epoch + 1$.

- Step 6: If mse^k satisfies the target value of the loss function mse_{ref}^k, the training process is completed, and capacity estimation and analysis should be conducted. Otherwise, if the maximum epoch E_{max} has arrived, adjust the parameters and redo the procedure.

4.4 CAPACITY ESTIMATION RESULTS

4.4.1 Experimental setup

Experiments are conducted to demonstrate the effects of the proposed algorithm. The dataset with the labels of the ten EVs shown in Figure 4.2 is applied and divided into a training dataset, a validation dataset, and a testing dataset. The ten EVs contained a large amount of charging snippets in various DOC ranges, which are filtered into 5697 snippets, as shown in Table 4.1. The DOC histogram of the 5697 snippets is shown in Figure 4.10, which demonstrates that the dataset had high quality for neural network training, validation, and testing. Then, the 5697 snippets are divided into training, validation, and testing snippets in the ratio of 0.8:0.05:0.15.

With the partition of training, validation, and testing datasets, eight of the ten EVs (EV1–EV8) acted as the training dataset, almost half of EV9's data are selected as the validation dataset, and the other half of EV9 and EV10 acted as the testing dataset. The selection of this partition is done according to the data length of the ten EVs in the time sequence, because 80% of the 5697 snippets are from the battery data of EV1–EV8, and we tried to make all EVs separate from each other. A large part (80%) of the EV snippets are considered training data to offer us many diverse battery degradation types of EVs as possible during the training process. Then, the

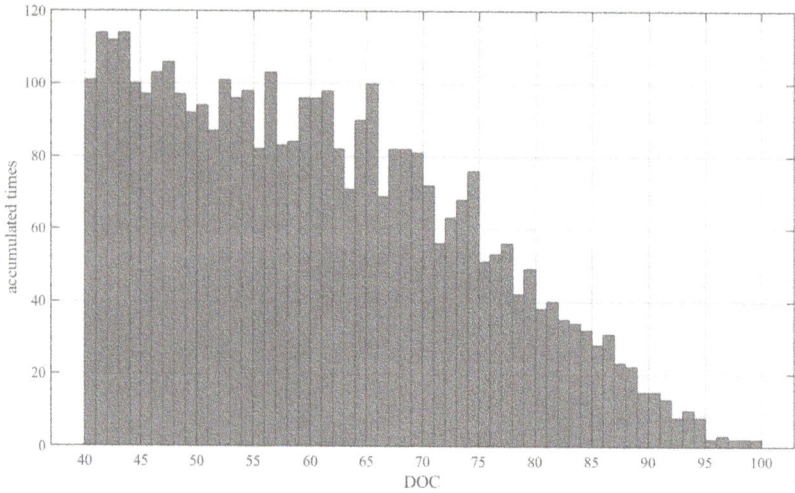

Figure 4.10 The histogram of the DOC of all inputs for all 5697 snippets.

algorithm could learn enough battery information for further testing estimation. Every snippet corresponds to a battery capacity value of an EV. Hence, the 15% testing part included 854 snippets of EVs, corresponding to 854 capacity points, which are enough for testing. The validation part only needed a small section of data for the gradient threshold, and thus the validation section only took up 5%. Considering the ICA part, the inputs of the proposed algorithm are selected, as shown in Table 4.2.

Table 4.2 The selected inputs of the proposed algorithm

No.	1	2	3	4	5	6
Type	average current	start SOC	end SOC	DOC	start voltage	end voltage
No.	7	8	9	10	11	12
Type	mileage	Ah quantity	start temperature	end temperature	peak1 magnitude	peak2 magnitude

The algorithm parameters are pre-tuned for training performance. The determined parameters of the proposed PIRNN are shown in Table 4.3. The main tuning parameters of the FOGD method are the learning rate l_r and the fractional order α.

4.4.2 Estimation results for battery degradation

We took the variables in Table 4.2 as the inputs, the parameters in Table 4.3 as the network parameters, and the capacity values as the output of the proposed algorithm; the estimation results are presented in this section. Figure 4.11 is the training process with the validation dataset, and the loss of the testing dataset is also shown in Figure 4.11. Since battery degradation is a key changing state with time, the training

Table 4.3 Algorithm parameters for training, validation, and testing

Name	Value	Name	Value
state delays in PIRNN	1:2	hidden layer size	8
performance function	MSE	maximum epoch	3000
train function	FOGD	train–validation–test	0.8:0.05:0.15
learning rate l_r	0.0001	training goal	1
fractional order α	0.8	validation times	50

dataset, validation dataset, and testing dataset are plotted in a "pseudo" time–series sequence for the learning algorithm. The loss function is the MSE of the outputs, that is, the capacity estimation values. Figure 4.11 demonstrates that the proposed algorithm could converge smoothly and rapidly with the fractional-order gradient method, and the loss could decrease to a small value (<10), less than $1/10$ of the value of the loss in the beginning (180 epochs), which shows that the proposed PIRNN algorithm could learn the battery aging information with the fractional-order gradient method.

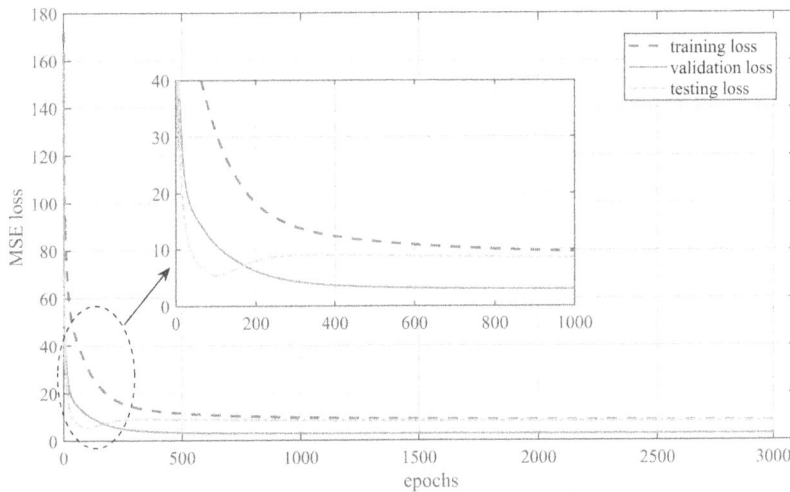

Figure 4.11 The training process of the proposed fractional-order recurrent neural network with the physics–informed method.

Figure 4.12 presents the estimation outputs of the proposed PIRNN algorithm, and the estimation results of the ten EVs are presented in a time series accompanied by the capacity labels shown in Figure 4.2. For comparison, the results of a conventional RNN with the integer-order GD method and ICA inputs are also presented in Figure 4.12. All parameters of the proposed PIRNN and conventional RNN with the GD method are the same except for the fractional order. In the results of Figure 4.12, since the dataset of the ten EVs is divided into training–validation–testing parts by the ratio of as 0.8:0.05:0.15, eight of the ten EVs (EV1–EV8) act as the training

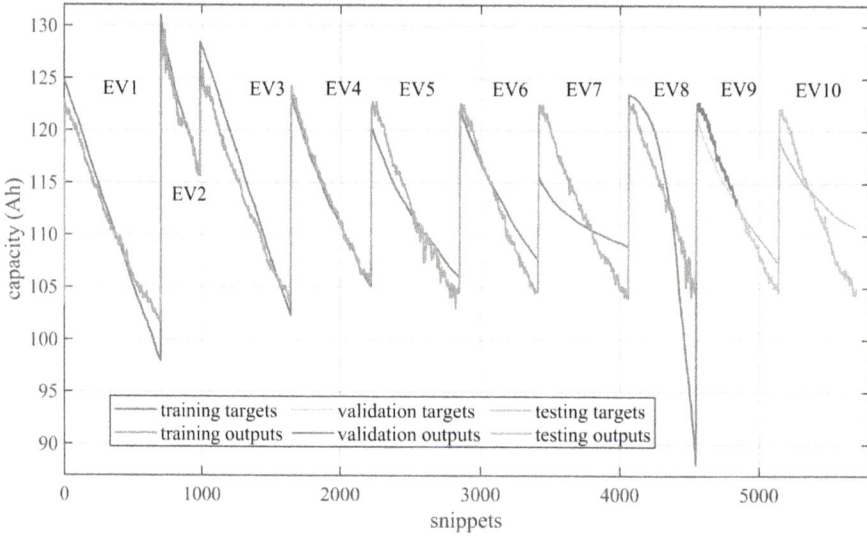

(a) PIRNN with FOGD and ICA inputs

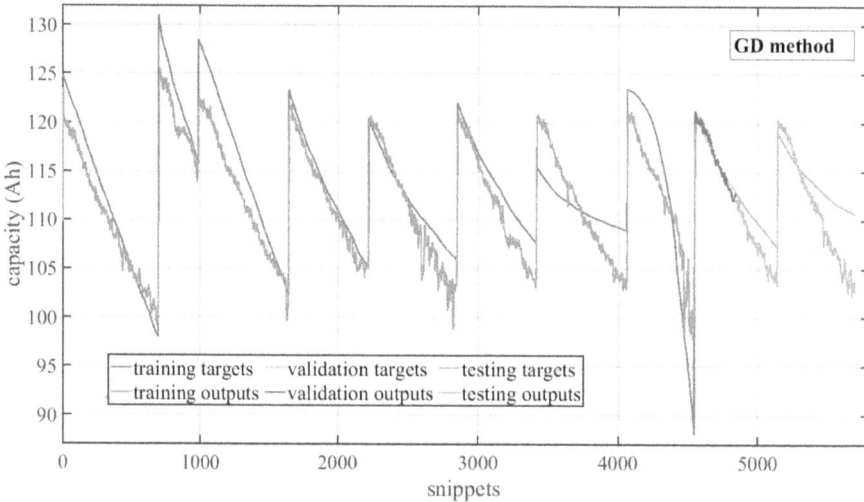

(b) RNN with GD and ICA inputs

Figure 4.12 The capacity estimation results of the proposed fractional-order recurrent neural network with the physics–informed method, compared to a conventional RNN with the integer-order GD method.

dataset, almost half of EV9's data is selected as the validation dataset, and the other half of EV9 and EV10 act as the testing dataset. Generally, the estimation results can fit most of the capacity of the ten EVs and track the degradation trend of the LFP battery.

It should be noted that the transient variation in the shift points between every two EVs should not be taken as the estimation results of the proposed PIRNN algorithm. For example, the shift point between EV1 and EV2 caused a large capacity change from 98 Ah to 130 Ah, resulting a large amount of noise for the estimation output of the PIRNN in the beginning of EV2. Every point shown in Figure 4.12 presents a snippet among the 5697 snippets of the ten EVs, and the time range of every EV curve covers from the year 2018 to 2022, which is already a long running period. Although it still seems that the estimation performance of the algorithm in Figure 4.12 does not hold very well in every part of the capacity degradation curve, the estimation results can remain stable in most of the battery lifetime. Specifically, the training performance on EV7 and EV8 is the worst among the ten EVs, which may have been caused by the capacity change rates of EV7 and EV8, different from the other six training EVs. This also influenced the performance on the testing dataset, as shown in the right region of Figure 4.12, and the testing outputs turned out to be similar to those of the first six EVs.

To further illustrate the performance, the estimation errors (output labels) and the relative errors are also provided in Figures 4.13 and 4.14, respectively. The relative errors are calculated by the absolute values of the output labels/labels, so the relative errors are calculated based on the battery aging labels rather than the rated capacity, which is a real-time error index.

Generally, the errors and relative errors in Figures 4.13 and 4.14 demonstrate that the proposed algorithm could learn the battery's physics–informed knowledge and control the errors within an acceptable level when the dataset holds a similar degradation trend. The errors of eight of the ten EVs shown in Figure 4.13 are lower than 5Ah, which is small enough for a battery pack with a 130Ah rated capacity.

First of all, it should be noted that the shift points between every two EVs bring large and transient changes in the error values, such as the shift error between EV1 and EV2, EV3 and EV4, or EV8 and EV9, and the transient change in error should be considered noise rather than the estimation output. On the other hand, the error curves with shift points also prove that the proposed PIRNN can hold enough robustness with initial data noise. As shown in Figure 4.14, the relative errors of the first six EVs (EV1–EV6) and EV9 during the whole battery lifetime could be lower than 3%, which is also good enough for capacity estimation. However, in the end of the capacity curves of the ten EVs, especially EV5, EV6, and EV10, the proposed algorithm cannot learn the capacity trends as well as in the early stage, because the time interval between two snippets in this stage may be more than one month, and the EV battery may near the end of its lifetime, which makes it harder for algorithm to learn the trends. Moreover, EV7 and EV8 have batteries with faster degradation trends, and EV8 nearly has a dramatic drop in the end of its lifetime, so the proposed algorithm cannot catch the sudden drops by its current FOGD method.

Figures 4.12–4.14 put the estimation results of the ten EVs together, which aims to show the general effectiveness of the proposed algorithm. To better illustrate the details, we extracted single estimation results for every EV from the ten EVs. Figure 4.15 shows the detailed results of EV4 in the training dataset, and Figure 4.16 shows the comparison of EV6 in the training dataset and EV10 in the testing dataset.

In Figures 4.15 and 4.16, both capacity estimation outputs and the corresponding relative errors are provided, and the x–axes of the figures are the snippet numbers in all the 5697 obtained snippets of the ten EVs, so they do not start from 1. The results

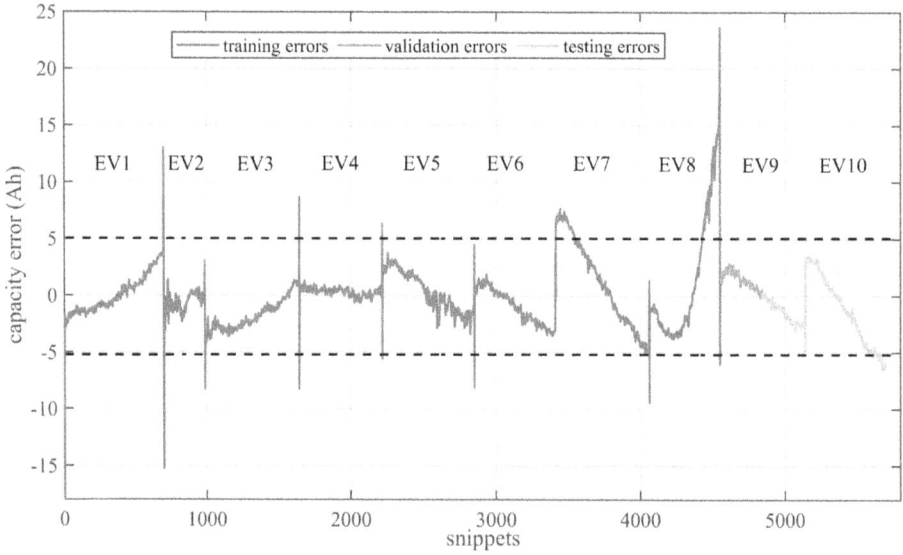

(a) PIRNN with FOGD and ICA inputs

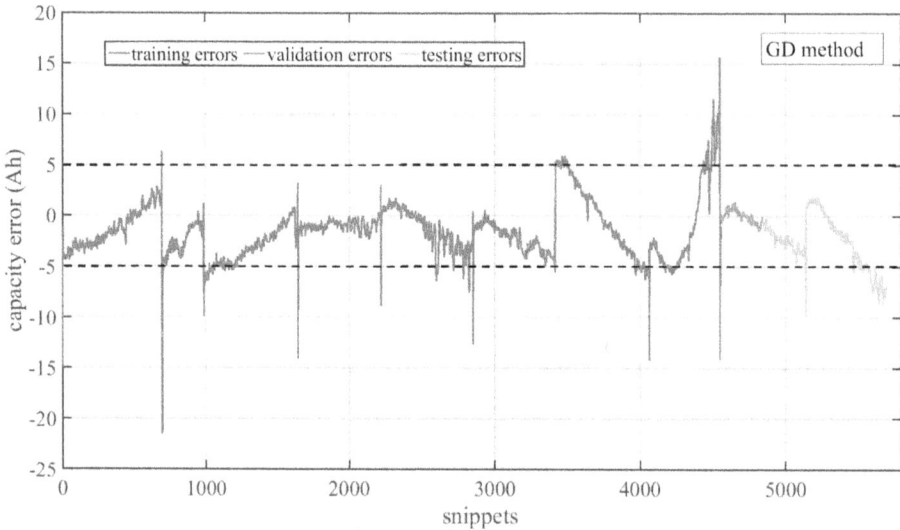

(b) RNN with GD and ICA inputs

Figure 4.13 The estimation errors of the proposed fractional-order recurrent neural network with the physics-informed method, compared to a conventional RNN with the integer-order GD method.

of EV4 in Figure 4.15 show the best fitting of capacity labels among the ten EVs. This does not mean that there is an over-fitting of EV4 data, because by comparing Figure 4.15 with Figure 4.12, the estimated outputs of the other nine EVs can be adjusted by the algorithm and the estimated degradation trends of the other nine EVs vary from each other. Meanwhile, in Figure 4.16, it can be observed that the estimated

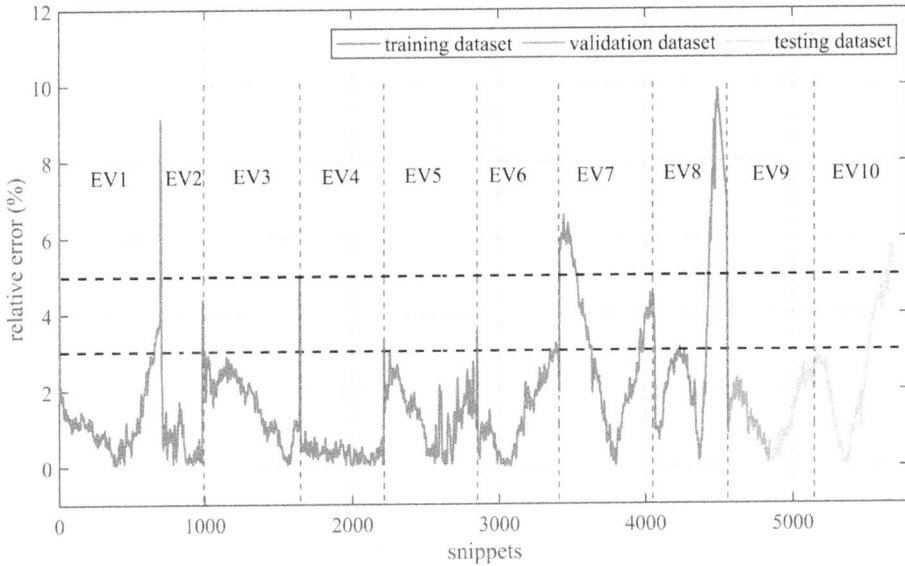

(a) PIRNN with FOGD and ICA inputs

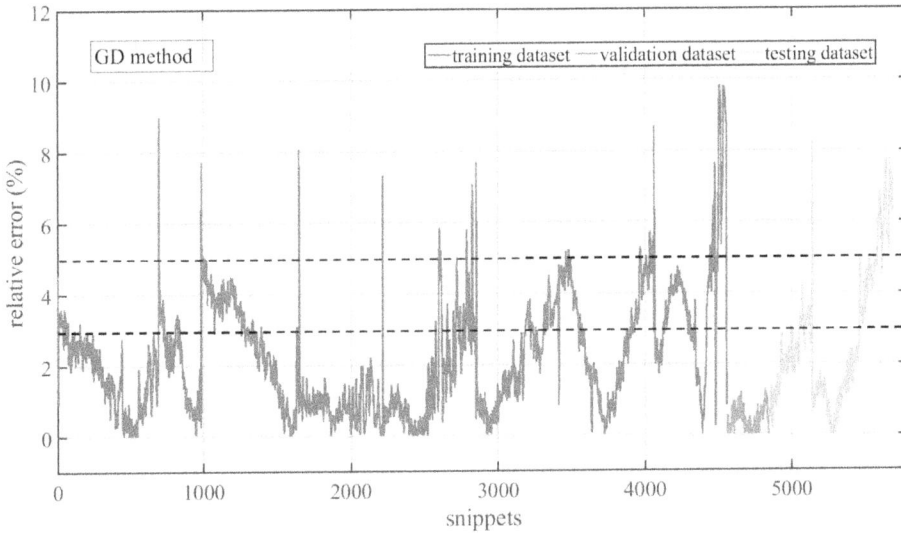

(b) RNN with GD and ICA inputs

Figure 4.14 The relative errors of the proposed fractional-order recurrent neural network with the physics-informed method, compared to a conventional RNN with the integer-order GD method.

results in the end stage of the battery pack are not as accurate as those in the early stage. Thus, we find the critical point when the relative error exceeds 3%, which is also marked in Figure 4.16. For EV6 in the training dataset, the critical point happened on 9 October 2020, and the capacity label is 110.334 Ah with the SOH = 84.9%. For EV10 in the testing dataset, the critical point happened on 1 October 2020,

(a) EV4 estimation

(b) relative error of EV4

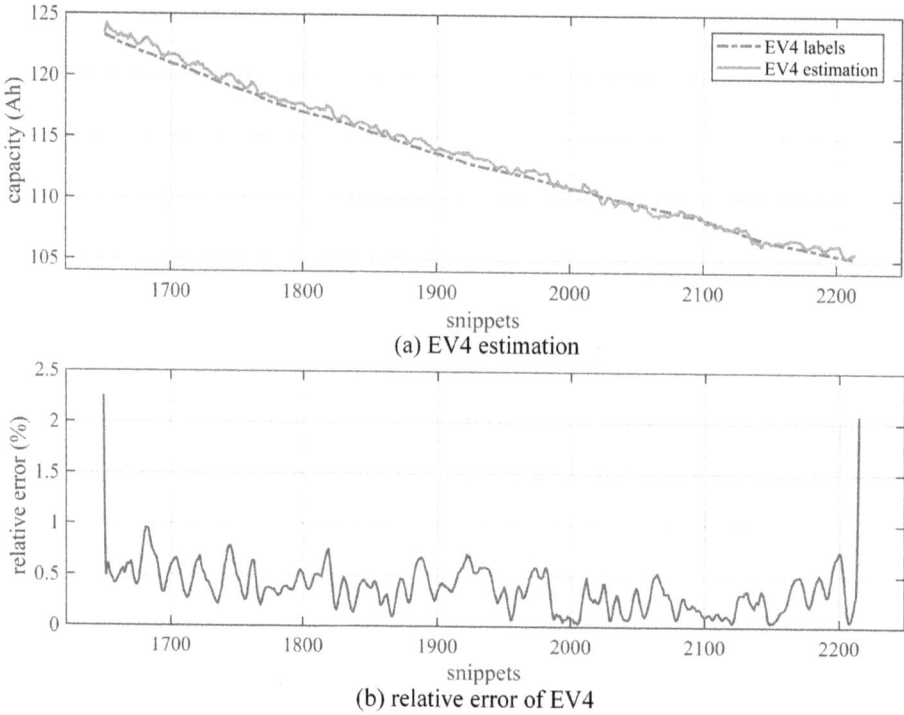

Figure 4.15 Detailed estimation and relative error of EV4 in training dataset. (a) EV4 estimation results with labels. (b) EV4 relative error.

and the capacity label is 112.179 Ah with the SOH $= 86.3\%$. It can be seen that the estimation errors stayed under 3% until the battery lifetime reached around 85%; thus, this proves that the proposed algorithm could hold the estimation accuracy over three-quarters of the battery lifetime.

In this section, we present the estimation results and analysis of the proposed PIRNN algorithm for fast battery degradation estimation. The results show that the proposed algorithm could achieve a stable relative error $< 3\%$ for most of the ten EVs over three-quarters of the battery lifetime. The convergence process also showed that the proposed algorithm could learn the battery information through the IC inputs and FOGD method. However, the estimation accuracy decreased at the end of the battery lifetime and should be improved by embedding further physical information of the battery. Moreover, the performance on the battery with a dramatic degradation rate should be improved by introducing other physical inner degradation variables. It should be noted that all ten EVs collected in the dataset have a battery with faster degradation rather than a power–law trend, which may already make it harder to conduct capacity estimation in realistic operation conditions.

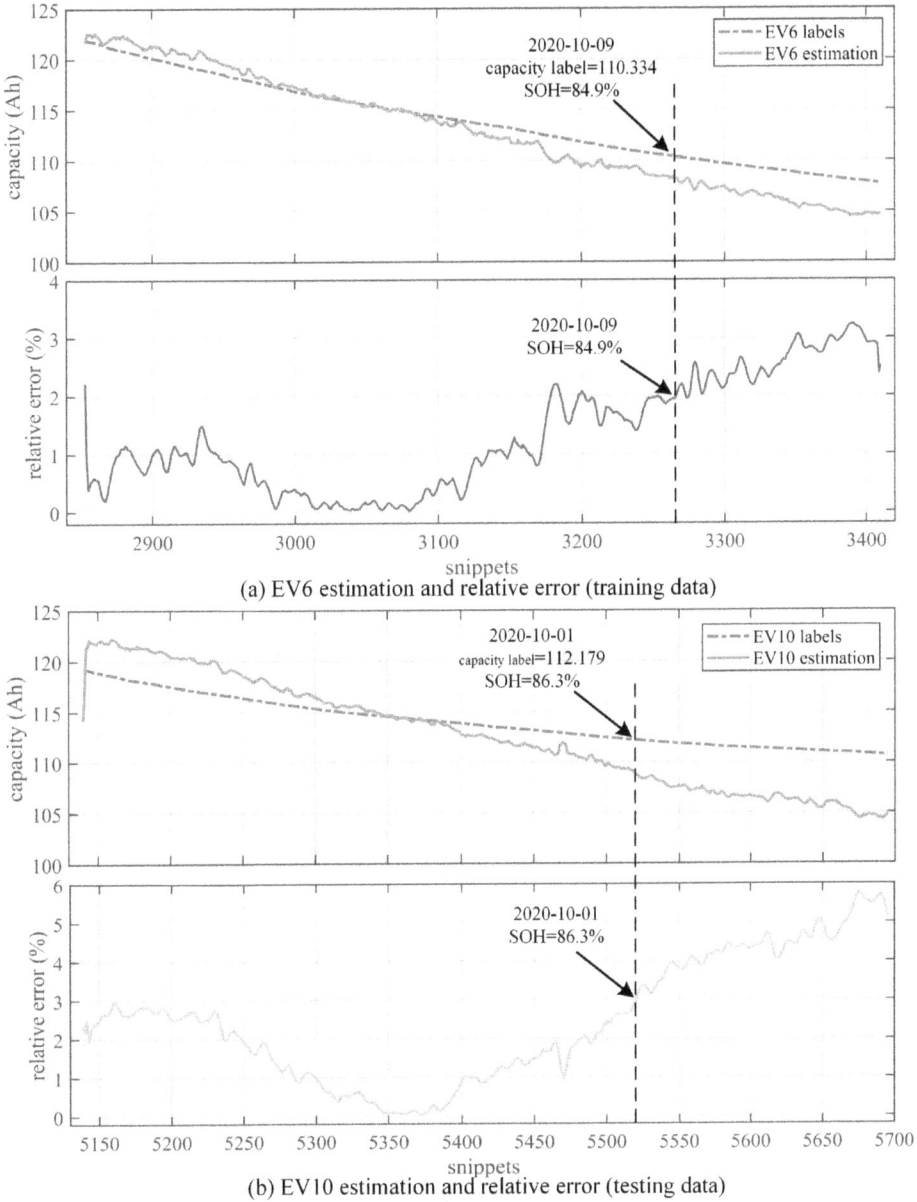

(a) EV6 estimation and relative error (training data)

(b) EV10 estimation and relative error (testing data)

Figure 4.16 Comparison of training results and testing results with EV6 and EV10. (a) EV6 estimation and relative error in training dataset. (b) EV10 estimation and relative error in testing data.

4.5 CHAPTER SUMMARY

This chapter proposes a physics-informed method for a fractional-order recurrent neural network to perform battery capacity estimation, especially for a battery with fast degradation. A characteristic extracted from ICA is first applied to enhance the learning of the battery's inner information and the algorithm interpretability.

The fractional–order gradient descent method is applied to accelerate the training process of the proposed algorithm. Hence, a fractional-order recurrent neural network with a physics-informed method is designed for battery capacity estimation. Specifically, the investigated battery dataset in this chapter is from the LFP batteries of ten EVs with fast degradation, which is different from typical power–law capacity aging. The proposed algorithm is directly verified on the realistic sampled data and could achieve a relative error < 3% for most of the ten EVs over three–quarters of the battery lifetime. The results show that the proposed algorithm could achieve stable performance in the degradation estimation when the batteries hold similar degradation trends. The results demonstrate that the research direction of embedding battery knowledge by the physics–informed method is worthy of investigation. However, additional inner physical or even electrochemical variables should be added to enhance the physics-informed degree of machine learning. Moreover, other types of machine learning methods can also be explored to achieve an even better performance in future investigations.

Bibliography

[1] Maher GM Abdolrasol, SM Suhail Hussain, Taha Selim Ustun, Mahidur R Sarker, Mahammad A Hannan, Ramizi Mohamed, Jamal Abd Ali, Saad Mekhilef, and Abdalrhman Milad. Artificial neural networks based optimization techniques: A review. *Electronics*, 10(21):2689, 2021.

[2] Ruslan Abdulkadirov, Pavel Lyakhov, and Nikolay Nagornov. Survey of optimization algorithms in modern neural networks. *Mathematics*, 11(11):2466, 2023.

[3] Mohd Ashraf Ahmad, Nik Mohd Zaitul Akmal Mustapha, Ahmad Nor Kasruddin Nasir, Mohd Zaidi Mohd Tumari, Raja Mohd Taufika Raja Ismail, and Zuwairie Ibrahim. Using normalized simultaneous perturbation stochastic approximation for stable convergence in model-free control scheme. In *2018 IEEE International Conference on Applied System Invention (ICASI)*, pages 935–938. IEEE, 2018.

[4] Shun-ichi Amari. Backpropagation and stochastic gradient descent method. *Neurocomputing*, 5(4-5):185–196, 1993.

[5] Mahshid N Amiri, Anne Haakansson, Odne S Burheim, and Jacob J Lamb. Lithium-ion battery digitalization: Combining physics-based models and machine learning. *Renewable and Sustainable Energy Reviews*, 200:114577, 2024.

[6] Yingjie Chen, Huaqin Zhang, Jichao Hong, Yankai Hou, Jingsong Yang, Chi Zhang, Shikun Ma, Xinyang Zhang, Haixu Yang, Fengwei Liang, et al. Lithium plating detection of lithium-ion batteries based on the improved variance entropy algorithm. *Energy*, 299:131574, 2024.

[7] Zheng Chen, Hongqian Zhao, Yuanjian Zhang, Shiquan Shen, Jiangwei Shen, and Yonggang Liu. State of health estimation for lithium-ion batteries based

on temperature prediction and gated recurrent unit neural network. *Journal of Power Sources*, 521:230892, 2022.

[8] Kaushik Das, Roushan Kumar, and Anurup Krishna. Analyzing electric vehicle battery health performance using supervised machine learning. *Renewable and Sustainable Energy Reviews*, 189:113967, 2024.

[9] Heba-Allah I El-Azab, RA Swief, Noha H El-Amary, and HK Temraz. Seasonal electric vehicle forecasting model based on machine learning and deep learning techniques. *Energy and AI*, 14:100285, 2023.

[10] Xuning Feng, Yu Merla, Caihao Weng, Minggao Ouyang, Xiangming He, Bor Yann Liaw, Shriram Santhanagopalan, Xuemin Li, Ping Liu, Languang Lu, et al. A reliable approach of differentiating discrete sampled-data for battery diagnosis. *ETransportation*, 3:100051, 2020.

[11] Dongxu Guo, Geng Yang, Xuning Feng, Xuebing Han, Languang Lu, and Minggao Ouyang. Physics-based fractional-order model with simplified solid phase diffusion of lithium-ion battery. *Journal of Energy Storage*, 30:101404, 2020.

[12] Dongxu Guo, Geng Yang, Xuebing Han, Xuning Feng, Languang Lu, and Minggao Ouyang. Parameter identification of fractional-order model with transfer learning for aging lithium-ion batteries. *International Journal of Energy Research*, 45(9):12825–12837, 2021.

[13] Xuebing Han, Minggao Ouyang, Languang Lu, Jianqiu Li, Yuejiu Zheng, and Zhe Li. A comparative study of commercial lithium ion battery cycle life in electrical vehicle: Aging mechanism identification. *Journal of Power Sources*, 251:38–54, 2014.

[14] Jichao Hong, Zhenpo Wang, Wen Chen, Leyi Wang, Peng Lin, and Changhui Qu. Online accurate state of health estimation for battery systems on real-world electric vehicles with variable driving conditions considered. *Journal of Cleaner Production*, 294:125814, 2021.

[15] Dickson NT How, MA Hannan, MS Hossain Lipu, and Pin Jern Ker. State of charge estimation for lithium-ion batteries using model-based and data-driven methods: A review. *IEEE Access*, 7:136116–136136, 2019.

[16] Taotao Hu, Zheng He, Xiaojun Zhang, and Shouming Zhong. Finite-time stability for fractional-order complex-valued neural networks with time delay. *Applied Mathematics and Computation*, 365:124715, 2020.

[17] Yujiao Huang, Xiaoyan Yuan, Haixia Long, Xinggang Fan, and Tiaoyang Cai. Multistability of fractional-order recurrent neural networks with discontinuous and nonmonotonic activation functions. *IEEE Access*, 7:116430–116437, 2019.

[18] Julakha Jahan Jui, Mohd Ashraf Ahmad, MM Imran Molla, and Muhammad Ikram Mohd Rashid. Optimal energy management strategies for hybrid electric vehicles: A recent survey of machine learning approaches. *Journal of Engineering Research*, 2024.

[19] George Em Karniadakis, Ioannis G Kevrekidis, Lu Lu, Paris Perdikaris, Sifan Wang, and Liu Yang. Physics-informed machine learning. *Nature Reviews Physics*, 3(6):422–440, 2021.

[20] Xin Lai, Wei Yi, Yifan Cui, Chao Qin, Xuebing Han, Tao Sun, Long Zhou, and Yuejiu Zheng. Capacity estimation of lithium-ion cells by combining model-based and data-driven methods based on a sequential extended Kalman filter. *Energy*, 216:119233, 2021.

[21] Weihan Li, Jiawei Zhang, Florian Ringbeck, Dominik Jóst, Lei Zhang, Zhongbao Wei, and Dirk Uwe Sauer. Physics-informed neural networks for electrode-level state estimation in lithium-ion batteries. *Journal of Power Sources*, 506:230034, 2021.

[22] MS Hossain Lipu, MS Abd Rahman, M Mansor, Tuhibur Rahman, Shaheer Ansari, Abu M Fuad, and MA Hannan. Data driven health and life prognosis management of supercapacitor and lithium-ion battery storage systems: Developments, implementation aspects, limitations, and future directions. *Journal of Energy Storage*, 98:113172, 2024.

[23] MS Hossain Lipu, MA Hannan, Aini Hussain, Afida Ayob, Mohamad HM Saad, Tahia F Karim, and Dickson NT How. Data-driven state of charge estimation of lithium-ion batteries: Algorithms, implementation factors, limitations and future trends. *Journal of Cleaner Production*, 277:124110, 2020.

[24] Tianle Liu, Promit Ghosal, Krishnakumar Balasubramanian, and Natesh Pillai. Towards understanding the dynamics of Gaussian-Stein variational gradient descent. *Advances in Neural Information Processing Systems*, 36, 2024.

[25] Weilbeer Marc. Efficient numerical methods for fractional differential equations and their analytical background. *Mathematics*, 2005.

[26] RenHao Mok and Mohd Ashraf Ahmad. Smoothed functional algorithm with norm-limited update vector for identification of continuous-time fractional-order Hammerstein models. *IETE Journal of Research*, 70(2):1814–1832, 2024.

[27] Achraf Nasser-Eddine, Benoît Huard, Jean-Denis Gabano, and Thierry Poinot. A two steps method for electrochemical impedance modeling using fractional order system in time and frequency domains. *Control Engineering Practice*, 86:96–104, 2019.

[28] Man-Fai Ng, Jin Zhao, Qingyu Yan, Gareth J Conduit, and Zhi Wei Seh. Predicting the state of charge and health of batteries using data-driven machine learning. *Nature Machine Intelligence*, 2(3):161–170, 2020.

[29] Darius Roman, Saurabh Saxena, Valentin Robu, Michael Pecht, and David Flynn. Machine learning pipeline for battery state-of-health estimation. *Nature Machine Intelligence*, 3(5):447–456, 2021.

[30] Markus Schindler, Johannes Sturm, Sebastian Ludwig, Julius Schmitt, and Andreas Jossen. Evolution of initial cell-to-cell variations during a three-year production cycle. *ETransportation*, 8:100102, 2021.

[31] Dian Sheng, Yiheng Wei, Yuquan Chen, and Yong Wang. Convolutional neural networks with fractional order gradient method. *Neurocomputing*, 408:42–50, 2020.

[32] Dandan Song, Zhe Gao, Haoyu Chai, and Zhiyuan Jiao. An adaptive fractional-order extended Kalman filtering approach for estimating state of charge of lithium-ion batteries. *Journal of Energy Storage*, 85:111089, 2024.

[33] Kenneth O Stanley, Jeff Clune, Joel Lehman, and Risto Miikkulainen. Designing neural networks through neuroevolution. *Nature Machine Intelligence*, 1(1):24–35, 2019.

[34] Laisuo Su, Mengchen Wu, Zhe Li, and Jianbo Zhang. Cycle life prediction of lithium-ion batteries based on data-driven methods. *ETransportation*, 10:100137, 2021.

[35] Valentin Sulzer, Peyman Mohtat, Antti Aitio, Suhak Lee, Yen T Yeh, Frank Steinbacher, Muhammad Umer Khan, Jang Woo Lee, Jason B Siegel, Anna G Stefanopoulou, et al. The challenge and opportunity of battery lifetime prediction from field data. *Joule*, 5(8):1934–1955, 2021.

[36] Jinpeng Tian, Rui Xiong, Jiahuan Lu, Cheng Chen, and Weixiang Shen. Battery state-of-charge estimation amid dynamic usage with physics-informed deep learning. *Energy Storage Materials*, 50:718–729, 2022.

[37] Jinpeng Tian, Rui Xiong, Weixiang Shen, Jiahuan Lu, and Xiao-Guang Yang. Deep neural network battery charging curve prediction using 30points collected in 10 min. *Joule*, 5(6):1521–1534, 2021.

[38] Baojin Wang, Zhiyuan Liu, Shengbo Eben Li, Scott Jason Moura, and Huei Peng. State-of-charge estimation for lithium-ion batteries based on a nonlinear fractional model. *IEEE Transactions on Control Systems Technology*, 25(1):3–11, 2016.

[39] Fujin Wang, Zhi Zhai, Zhibin Zhao, Yi Di, and Xuefeng Chen. Physics-informed neural network for lithium-ion battery degradation stable modeling and prognosis. *Nature Communications*, 15(1):4332, 2024.

[40] Xueyuan Wang, Xuezhe Wei, Jiangong Zhu, Haifeng Dai, Yuejiu Zheng, Xiaoming Xu, and Qijun Chen. A review of modeling, acquisition, and application of lithium-ion battery impedance for onboard battery management. *ETransportation*, 7:100093, 2021.

[41] Yuan Wang, Yangquan Chen, and Xiaozhong Liao. State-of-art survey of fractional order modeling and estimation methods for lithium-ion batteries. *Fractional Calculus and Applied Analysis*, 22(6):1449–1479, 2019.

[42] Yanan Wang, Xuebing Han, Dongxu Guo, Languang Lu, Yangquan Chen, and Minggao Ouyang. Physics-informed recurrent neural network with fractional-order gradients for state-of-charge estimation of lithium-ion battery. *IEEE Journal of Radio Frequency Identification*, 6:968–971, 2022.

[43] Yanan Wang, Xuebing Han, Languang Lu, Yangquan Chen, and Minggao Ouyang. Sensitivity of fractional-order recurrent neural network with encoded physics-informed battery knowledge. *Fractal and Fractional*, 6(11):640, 2022.

[44] Yujie Wang, Li Wang, Mince Li, and Zonghai Chen. A review of key issues for control and management in battery and ultra-capacitor hybrid energy storage systems. *ETransportation*, 4:100064, 2020.

[45] Hao Yang, Penglei Wang, Yabin An, Changli Shi, Xianzhong Sun, Kai Wang, Xiong Zhang, Tongzhen Wei, and Yanwei Ma. Remaining useful life prediction based on denoising technique and deep neural network for lithium-ion capacitors. *ETransportation*, 5:100078, 2020.

[46] Qiangxiang Zhai, Hongmin Jiang, Nengbing Long, Qiaoling Kang, Xianhe Meng, Mingjiong Zhou, Lijing Yan, and Tingli Ma. Machine learning for full lifecycle management of lithium-ion batteries. *Renewable and Sustainable Energy Reviews*, 202:114647, 2024.

[47] Jie Zhang, Bo Xiao, Geng Niu, Xuanzhi Xie, and Saixiang Wu. Joint estimation of state-of-charge and state-of-power for hybrid supercapacitors using fractional-order adaptive unscented Kalman filter. *Energy*, 294:130942, 2024.

[48] Qi Zhang, Yunlong Shang, Yan Li, Naxin Cui, Bin Duan, and Chenghui Zhang. A novel fractional variable-order equivalent circuit model and parameter identification of electric vehicle li-ion batteries. *ISA Transactions*, 2019.

[49] Siyuan Zhang and Linbo Xie. Grafting constructive algorithm in feedforward neural network learning. *Applied Intelligence*, 53(10):11553–11570, 2023.

[50] Jingyuan Zhao, Xuning Feng, Quanquan Pang, Michael Fowler, Yubo Lian, Minggao Ouyang, and Andrew F Burke. Battery safety: Machine learning-based prognostics. *Progress in Energy and Combustion Science*, 102:101142, 2024.

[51] Jiangong Zhu, Yixiu Wang, Yuan Huang, R Bhushan Gopaluni, Yankai Cao, Michael Heere, Martin J Mühlbauer, Liuda Mereacre, Haifeng Dai, Xinhua Liu, et al. Data-driven capacity estimation of commercial lithium-ion batteries from voltage relaxation. *Nature Communications*, 13(1):1–10, 2022.

[52] Changfu Zou, Xiaosong Hu, Satadru Dey, Lei Zhang, and Xiaolin Tang. Non-linear fractional-order estimator with guaranteed robustness and stability for lithium-ion batteries. *IEEE Transactions on Industrial Electronics*, 65(7):5951–5961, 2017.

CHAPTER 5

Intelligent Modeling with Smart Sensors for Battery

The state estimation, life prediction, and safety warning of LIBs always depend on external features, such as voltage, current, and temperature. However, for the electrochemical reactions inside battery, the generated internal gas and heat cannot immediately be detected to be related to those external features, and these external signatures are not detected by the battery management system (BMS) until the end of battery lifespan or serious incidents (fire and explosion). Hence, besides fractional-order intelligent modeling for battery, intelligent modeling with smart sensors is also discussed in this chapter. This chapter aims at introducing more sensors to infer more internal signals for BMS, which is called smart perception in this chapter. First, smart perception is explained, and several types of battery sensors are presented in detail. Then, a kind of radio-frequency-based sensor called Walabot, known as microwave radar array, is introduced as an example of battery sensor application. Finally, a convolutional neural network (CNN) is designed to process Walabot signals for battery, and the classification results of battery voltage are provided to illustrate the effectiveness of battery sensors.

5.1   SMART PERCEPTION

Smart perception for battery means that internal reactions and states of battery could be inferred through additional sensors with useful signals. Hence, various battery sensors are investigated by researchers in recent years [4, 19, 17]. Despite the development in design of large amount of sensors in the field of materials chemistry, this chapter focuses on the foundational introduction of the main types of sensors for BMS. Generally, smart perception of battery states requires additional sensing signals, such as inner temperature, electrode potential, pressure, gas, and acoustics [15]. These types of sensors are introduced in the following sections.

DOI: 10.1201/9781003670902-5

5.1.1 Battery sensors

For smart perception, LIBs need to be assembled with more sensors to create useful signals in multi-aspects, such as electrical aspect (potential) [21], mechanical aspect (pressure) [6], thermal aspect (temperature) [18], and physical aspect (gas and sound) [14, 8, 12]. Different kinds of sensors may have their unique characteristics, and here we introduce several mainstream sensors as follows.

- Potential sensor: It is also called reference electrode, which is often employed to measure the liquid-phase potential of batteries. By implanting a reference electrode into the battery, it serves as an in-situ diagnostic tool, essentially functioning as an implanted potential sensor [21]. The implantation of reference electrodes enables decoupled analysis of cathodes and anodes, facilitating investigations into interfacial reactions between electrodes and electrolytes. This approach allows non-destructive monitoring of battery operating status without disassembly. Due to short-circuit risks and the confined working space, the practical implementation of reference electrodes requires careful consideration of electrolyte chemical composition and electrode geometry design.

- Pressure sensor: The breathing effect of batteries leads to variations in internal pressure within individual cells with state of charge (SOC), resulting in volume changes of active materials during lithium (de)intercalation [13, 20]. Such internal compressive pressure cannot be completely eliminated and will inevitably impact battery output performance and service life through charge-discharge cycles [9, 1]. Current research predominantly focuses on external pressure sensing devices, which faces challenges in timely detection of internal pressure variations. The mapping relationships between pressure-induced strain and battery safety status, charge-discharge states, and degree of degradation are under investigation, though the inherent stability of such systems still needs improvement.

- Temperature sensor: The temperature sensing signal is integrated into BMS to enable real-time monitoring of thermal status. However, the rapid cooling design of the flow channels in battery module introduces measurement errors between the channel temperature and the actual battery temperature, particularly for thermal runaway events. Researchers have explored implantable temperature sensor technologies, such as flexible electronics, to achieve in-situ multi-point temperature monitoring within battery cells [5].

- Gas sensor: During the cycling of lithium-ion batteries, gaseous species such as C_2H_4, H_2, and CO_2 are released. Elucidating gas generation mechanisms and implementing gas monitoring could enable the identification of specific electrochemical reactions and operational states [11]. For both battery cells and modules, physical signals involved in aging or thermal runaway include voltage, temperature, gas concentration, smoke emission, insulation resistance, mechanical stress, and gas pressure. However, the technical challenge of integrating gas sensors within the structurally sealed interior of individual battery cells remains largely unresolved.

- Acoustic sensor: This kind of sensor includes ultrasonic sensing [10, 16], microwave sensing, and sound sensing [12], which all detect battery and collect signals by emitting various types of waveforms. For example, sound sensing detects sound waves via piezoelectric materials, converting mechanical vibrations into electrical signals [12]. Acoustic sensors enable non-invasive monitoring of LIBs degradation by detecting stress wave emissions during electrochemical processes, such as mechanical stresses from SEI growth, lithium plating, or electrode cracking. Multi-sensor arrays spatially resolve internal defects via acoustic tomography. Validated applications include dendrite detection and electrolyte drying monitoring.

The comparison and performance of these five types of sensors are listed in Table 5.1. It compares the five types of sensors from seven aspects: accuracy, response range, compatibility, and service life are the performance indexes; state estimation, life prediction, and safety warning are the functions of the five types of sensors; finally the cost is listed in three levels (low, medium, and high).

Table 5.1 Performance of five types of sensors

Index	Potential	Pressure	Temperature	Gas	Acoustic
Accuracy	fair	fair	superior	superior	fair
Response range	superior	superior	superior	fair	fair
Compatibility	fair	superior	superior	inferior	inferior
Service life	inferior	fair	superior	fair	fair
State estimation	fair	fair	inferior	fair	inferior
Life prediction	superior	fair	superior	fair	fair
Safety warning	fair	fair	superior	superior	superior
Cost	medium	medium	low	high	high

For battery perception in the future, multifunctional integrated sensor is required with advantages of robustness, low barrier, and non-destructive implantation. In hardware aspect, material design and function optimization should be conducted to support further signal processing and wireless transfer technology. In software aspect, the accurate mapping of sensor signals to battery states should be investigated. Combination of coupled mechanical, thermal, and electrical signals are also necessary to achieve cooperative balance control for optimal battery power and long lifespan.

5.1.2 Walabot sensor

Walabot is a 3D (three-dimensional) radio frequency (RF) sensor developed by Vayyar Imaging, as shown in Figure 5.1. It is designed primarily for 3D imaging of internal structures in walls within the construction industry, such as detecting wooden pipes, metal pipes, electrical wires, and moving objects inside walls. It has also seen a few applications in the healthcare sector, including patient fall detection and breathing monitoring. However, since Walabot enables the detection of various objects along with its flexible development environment and API integrations, developers have adapted Walabot for a wide range of additional scenarios, including spatial object recognition and positioning, motion speed detection and tracking, gesture/movement sensing, and so on.

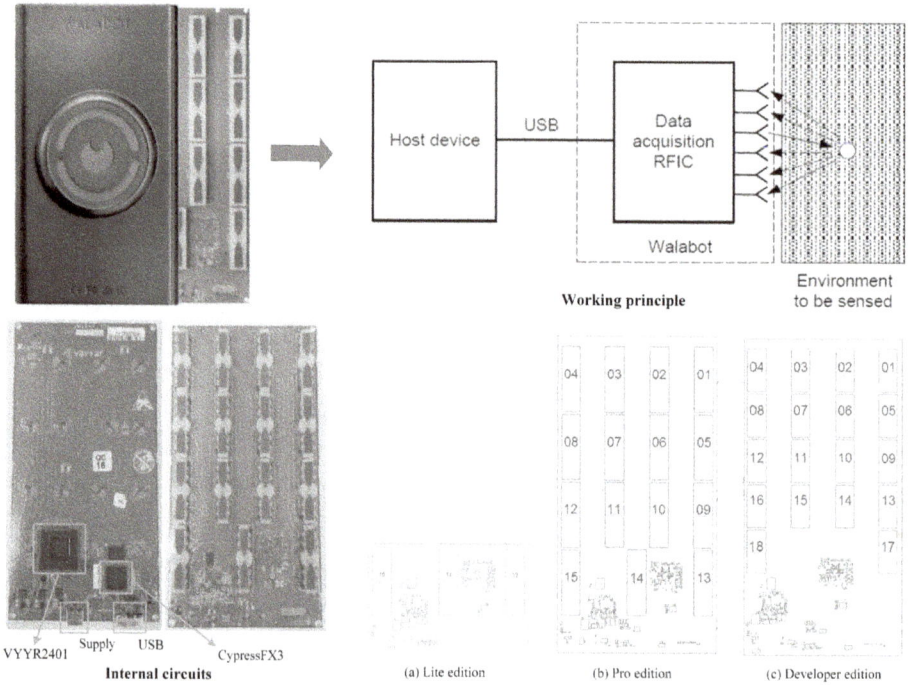

Figure 5.1 Walabot and its working principle.

The internal circuit structure and the working principle of Walabot are also shown in Figure 5.1. It mainly contains a transmitting antenna array, a receiving/processing chip VYYR2401, a communication chip CypressFX3, a power supply port, and a USB port. This RF sensor emits radio waves at specific frequencies through its antenna array on the circuit board. The VYYR2401 chip then receives and records the reflected waveforms. Finally, the CypressFX3 chip processes the recorded data, converting it into readable txt files or communicating with a host computer to store raw image data.

Vayyar Imaging offers three versions of Walabot, as shown in Figure 5.1, and their difference is the number of antenna arrays. We employ the developer edition, which offers the most features and highest flexibility. The developer edition operates with a radio wave transmission frequency range of 3.3 GHz to 10 GHz. The Walabot has its PC software and an APP and is compatible with Windows, Linux, and Raspberry Pi3 systems. Additionally, the data or images captured by Walabot reflect the waveforms of reflected signals, which contain information about the target object, such as distance, material composition, density, velocity, and so on. Based on its working principle, Walabot can collect distribution information within its effective range through emitted radio waves. Hence, Walabot can not only detect walls or moving objects, but also monitor internal distribution information of LIBs with battery parameters, such as voltage, real-time SOC, and internal resistance.

5.2 CONVOLUTIONAL NEURAL NETWORK WITH WALABOT

Considering that LIBs are electrochemical components, the whole internal chemical reactions affect the states of lithium ions and electrolytes when LIB is charged or discharged. By combining Walabot with machine learning to capture these characteristic states of lithium ions and electrolytes, it becomes possible to predict the corresponding voltage values of the lithium battery. Hence, we proposed a convolutional neural network with Walabot for voltage classification of battery in this section, to illustrate the application of Walabot with machine learning.

5.2.1 Battery images by Walabot

The Walabot with its PC software on Windows system can directly display images of the target object by analyzing its internal material composition relative to the angle and distance from the sensor. As shown in Figure 5.2, the left panel presents the lithium battery image obtained through Walabot detection, while the right panel shows its spatial position in a 3D coordinate system. This subsection focuses on collecting lithium battery images at different voltage levels from the left panel interface and verifying whether these images can correlate the corresponding voltage information.

The Walabot-generated images of lithium batteries in Figure 5.2 resemble photographs, while their classification via neural networks is analogous to facial

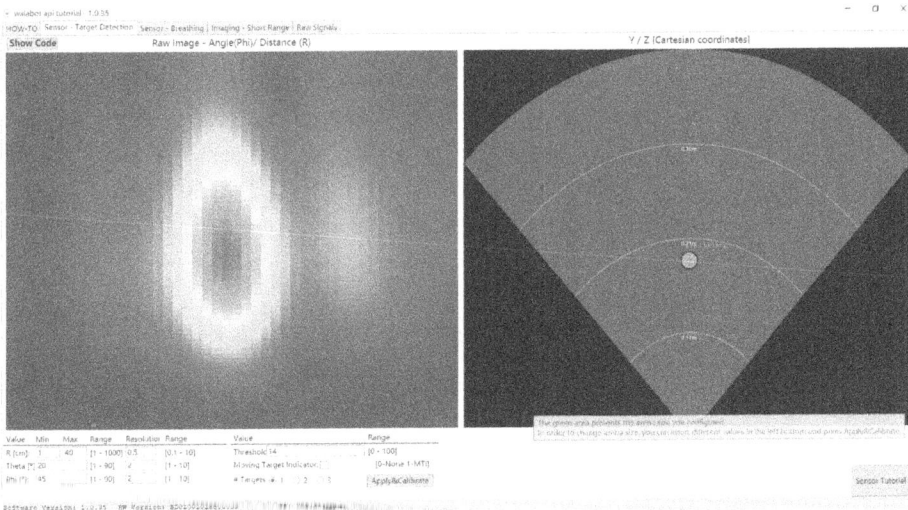

Figure 5.2 Battery image detected by Walabot and shown in its PC software on Windows system.

recognition methods. Thus, voltage-representative images captured by Walabot are termed "lithium battery face" images, with CNN selected as the machine learning approach for this "face" recognition task.

5.2.2 Convolutional neural network

Convolutional neural networks (CNNs) were first proposed by LeCun for image processing, which has two characteristics, i.e. spatially shared weights and spatial pooling [7]. A CNN can take in an input image, assign importance (learnable weights and biases) to various objects in the image and differentiate one from the other. The pre-processing effort required in a CNN is much lower as compared to other classification algorithms. While in primitive methods filters are hand-engineered, with enough training, CNNs have the ability to learn these characteristics. A typical architecture of a CNN is shown in Figure 5.3. It can be seen that CNNs exhibit distinct deep architectures compared to BPNNs or RNNs. A basic CNN typically comprises an input layer, convolutional layers, pooling layers, fully-connected layers, and an output layer. This structural complexity, characterized by multiple layers and intricate configurations, necessitates a higher parameter count.

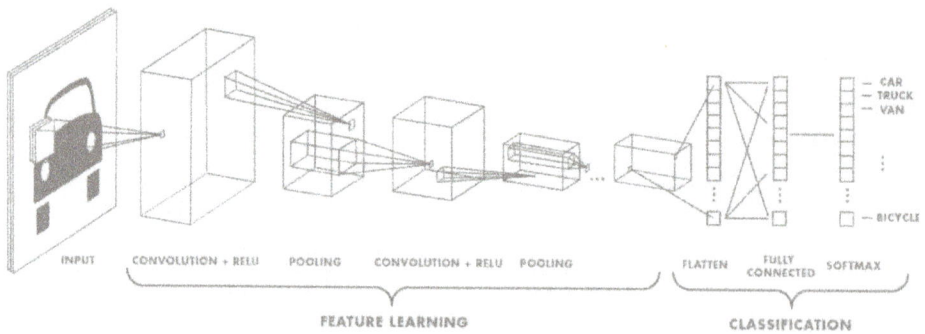

Figure 5.3 Typical architecture of a CNN.

CNN has been chosen because the very first CNN was actually designed just for imaging processing. If the "battery faces" of LIBs including information of electrolyte states and voltage values can be obtained by Walabot, then CNN should be efficient to process these "faces". Moreover, the data-driven characteristic of CNN makes it possible to conduct mapping from the collected feature images to the associated LIBs voltages, without any specific models or any other complex measurements.

5.3 BATTERY VOLTAGE CLASSIFICATION WITH WALABOT

5.3.1 Experimental setup

Experiments have been conducted to verify the effects of CNN to construct "battery faces" of LIBs voltage. Figure 5.4 shows the experiment setup. The parameters of five 3DR Iris+ 5100mAh/11.1V lithium polymer (LiPo) batteries are listed in Table 5.2, and the five LIBs are charged or discharged by the battery charger i–Charger–206B.

Images from Walabot are collected by the WalabotAPI software in Windows as shown in Figure 5.2.

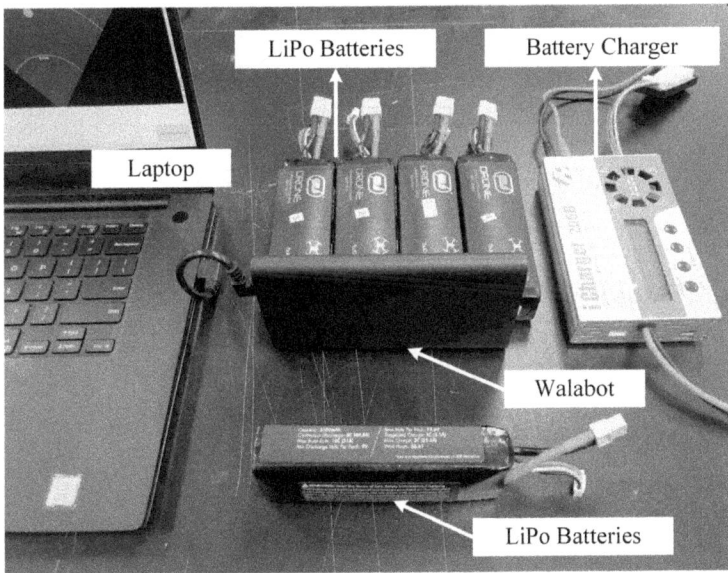

Figure 5.4 Experiment setup of wireless battery voltage classification using Walabot and machine learning.

Table 5.2 Parameters of 3DR IRIS+ 3S 5100Ah LiPo Battery

Name	Value	Name	Value
C rate	8C	Charge rate	1C(5.1A)
Voltage	11.1V	Capacity	5100Ah
Max voltage per pack	12.6V	Cell count	3S
Min voltage per pack	9V	Max charge rate	5C
Continuous discharge	8C	Max burst rate	10C

The five LiPo batteries were numbered as "No. 1, 2, 3, 4, 5", and all charged to full capacity firstly. As the capacity of every battery is vanishing gradually after usage, the five LiPo batteries have varied voltage values corresponding to the full charged capacity. The five LiPo batteries ("No.1, 2, 3, 4, 5") are 12.24V, 12.25V, 12.23V, 12.26V, and 12.33V when fully charged, separately. The aim of the proposed method is to classify voltage values of LIBs, so the fully charged states of the five LiPo batteries are considered as a class of voltage values, that is, 12.3V class. Similarly, other three classes of voltage values are chosen as 12V, 11V, 10V according to the parameters listed in Table 5.2. Hence, totally four classes of voltage were chosen in this experiment for voltage classification. After discharged from fully charged states by the battery charger i–Charger–206B, all five LiPo batteries rested 20 mins before voltage measurements. The experimental flow chart is shown in Figure 5.5.

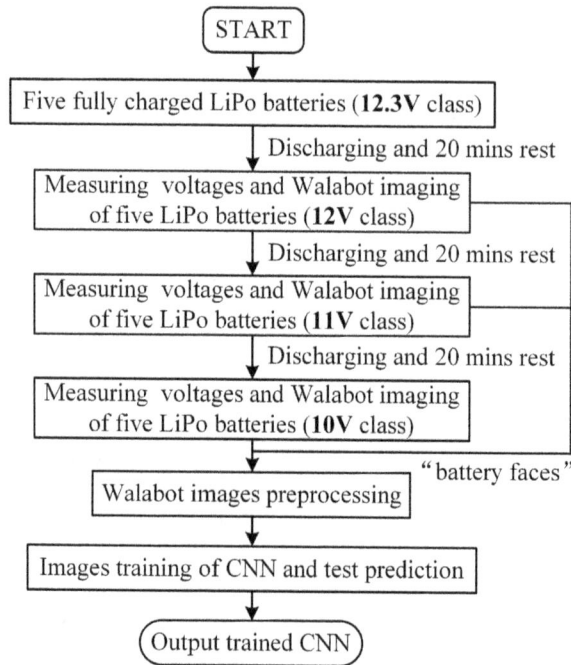

Figure 5.5 Experimental flow chart of voltage classification of LIBs by Walabot and CNN.

5.3.2 Voltage classification results

Following the flow chart shown in Figure 5.5, twenty images of five LiPo batteries have been collected as shown in Figure 5.6. It needs to be noted that all the "battery face" images come from screenshot rather than raw images due to the unavailable raw image save function of the Walabot software. Obviously, the original images are RGB images and the amount is relatively small for CNN training. Hence, image enhancement must be conducted for more extended images and further application, in this way, gray image and Wavelet transform are applied in this section.

After image enhancement, total 40 "battery face" images with labels were randomly split into 32 training images and 8 testing images. A basic CNN including an image input layer, a convolution layer, a pooling layer, and a fully connection layer was implemented based on the *trainNetwork* function in MATLAB. The image input layer is $766 \times 943 \times 1$ pixels size; the kernel size of convolution layer is 2×2, with 7 kernels; the size of pooling layer is 2×2, and has "strides" as 2; and the fully connection layer has four output according to the four labeled classes. The experiment result is presented in Figure 5.7, in which the max epoch was chosen as 100.

From Figure 5.7, it can be seen that when epochs reach 80, the mini-batch accuracy reaches 93.75%, after that, the accuracy changes between 81.25% and 93.75%. Hence, the experiment result in Figure 5.7 shows that the trained CNN has high batch accuracy based on the limited training images. As for the predict accuracy using the rest 8 testing images, it is not accurate enough due to the few sampling images from the five LiPo batteries by Walabot. However, this section presents the idea

| battery1 in 10.18V.PNG | battery1 in 11.14V.PNG | battery1 in 11.94V.PNG | battery1 in 12.24V.png |

| battery2 in 10.24V.PNG | battery2 in 11.13V.PNG | battery2 in 11.94V.PNG | battery2 in 12.25V.png |

| battery3 in 10.16V.PNG | battery3 in 11.06V.PNG | battery3 in 11.94V.PNG | battery3 in 12.23V.png |

| battery4 in 10.18V.PNG | battery4 in 11.05V.PNG | battery4 in 11.96V.PNG | battery4 in 12.26V.png |

| battery5 in 10.03V.PNG | battery5 in 11.21V.PNG | battery5 in 12.01V.PNG | battery5 in 12.33V.png |

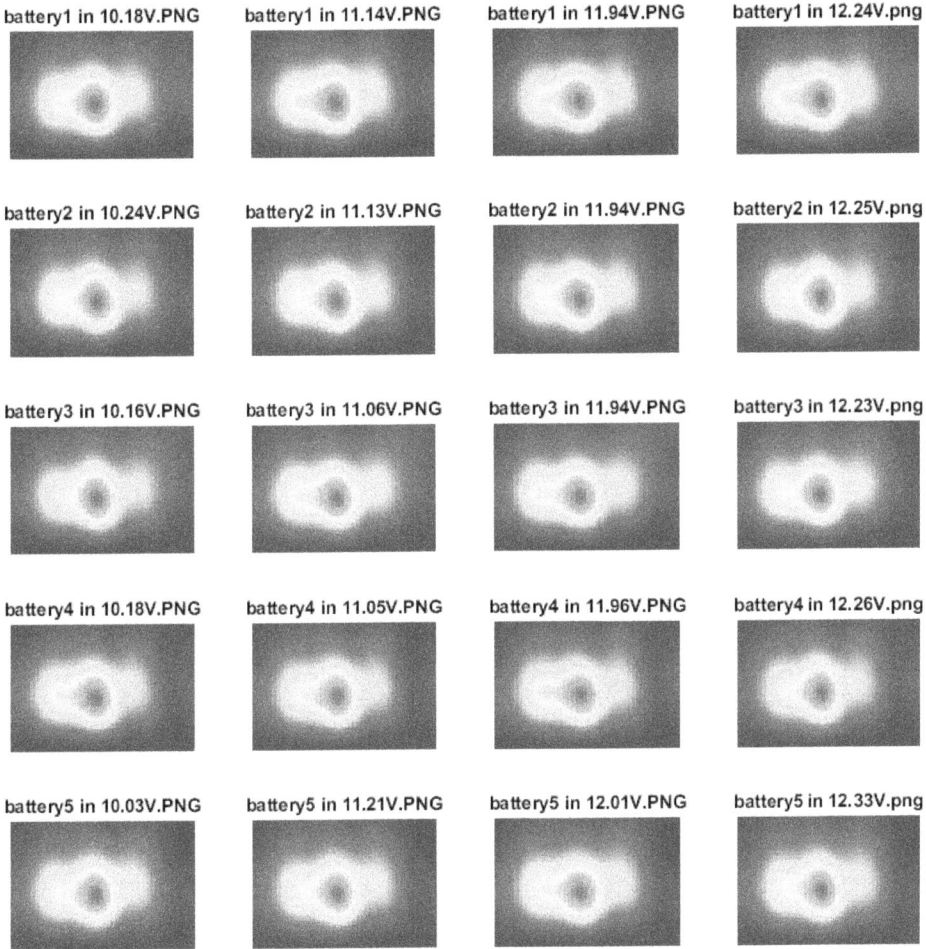

Figure 5.6   20 original RGB battery face images collected by Walabot.

of "battery faces" by Walabot and CNN, and a contactless way to conduct voltage classification of LIBs is proposed without any complicated modeling or parameter identification. Direct "battery face" imaging is a new way to characterize LIBs, and further work will be conducted for better performance.

5.4   BATTERY CAPACITY ESTIMATION BY MM-WAVE SENSOR

This section provide another example of sensor application for battery estimation [2]. A mm-wave sensor called ImageVK-74 from Vayyar is introduced with machine learning algorithms for battery capacity estimation.

5.4.1   Millimeter-wave sensor and its sensing system

The ImageVK-74 mm-wave sensor is from Vayyar designed and manufactured by mini-circuits. It has 20 transmitter (Tx) and 20 receiver (Rx) antennas (shown in Figure 5.8(a)) that can be configured to transmit and receive signals in the 62–69

Epoch	Iteration	Time Elapsed (seconds)	Mini-batch Loss	Mini-batch Accuracy	Base Learning Rate
1	1	10.38	1.7816	37.50%	1.00e-04
4	4	41.32	1.6431	34.38%	1.00e-04
8	8	82.59	1.4391	28.13%	1.00e-04
12	12	123.83	1.3499	31.25%	1.00e-04
16	16	164.97	1.2199	46.88%	1.00e-04
20	20	206.09	1.1033	56.25%	1.00e-04
24	24	247.14	0.9792	65.63%	1.00e-04
28	28	288.30	0.8023	68.75%	1.00e-04
32	32	329.36	0.7704	65.63%	1.00e-04
36	36	370.43	0.6319	68.75%	1.00e-04
40	40	411.62	2.2340	71.88%	1.00e-04
44	44	452.67	1.3236	34.38%	1.00e-04
48	48	493.83	1.2774	46.88%	1.00e-04
52	52	534.98	1.0924	56.25%	1.00e-04
56	56	576.10	0.9181	59.38%	1.00e-04
60	60	617.19	0.9094	65.63%	1.00e-04
64	64	658.19	0.8650	59.38%	1.00e-04
68	68	699.24	0.6695	62.50%	1.00e-04
72	72	740.42	0.4277	81.25%	1.00e-04
76	76	781.41	0.3603	81.25%	1.00e-04
80	80	822.58	0.1747	93.75%	1.00e-04
84	84	863.57	0.5909	87.50%	1.00e-04
88	88	904.72	0.6203	90.63%	1.00e-04
92	92	945.81	1.5850	87.50%	1.00e-04
96	96	986.83	0.6443	93.75%	1.00e-04
100	100	1027.99	0.8956	81.25%	1.00e-04

Figure 5.7 The CNN training result of "battery face" images related to voltage classification.

GHz frequency band. The high-resolution profile's 20 Tx and 20 Rx antennas and 400 raw signals make it perfect for high-resolution imaging and battery information collection. The ImageVK-74 millimeter wave sensor evaluation kit (EVK) has the whole millimeter-wave front-end and analog baseband signal chain for up to 40 antenna paths. A high-performance synthesizer produces a frequency-stepped continuous wave (CW) and sweeps over transmitting frequencies and antennas. Moreover, the Rx resolution bandwidth, start and stop frequency points, and the number of frequency points are customizable.

ImageVK-74 is used to construct a millimeter wave sensing system to collect battery signals. The sensing module consists of three major parts: the mm-wave transmitter and receiver (ImageVK-74), the Raspberry Pi 4B, and the visualization interface shown in Figure 5.8(b). The battery sample was placed under the direct mm-waves radar sensor, less than 2 cm. It can be noted that this sensing system can run on edge devices and work as an on-board module, which may be embedded into BMS. With the ImageVK-74 mm-waves sensor and its millimeter wave sensing system, battery features are collected from mm-wave reflectance and battery capacity is further indicated with machine learning algorithm.

The estimated battery comes from drones, which may need portable and contactless mobile device to evaluate their power batteries. The drone battery used is a Li-Po battery in 6S1P (6 cells in series connection without parallel connection). The fully charged voltage is 25.1 V (Open Circuit Voltage), and the capacity is 0500 mAh (as shown in Figure 5.8(b) right). To further capture the variation of the battery capacity during charging and discharging, we need to use an equivalent discharging circuit to

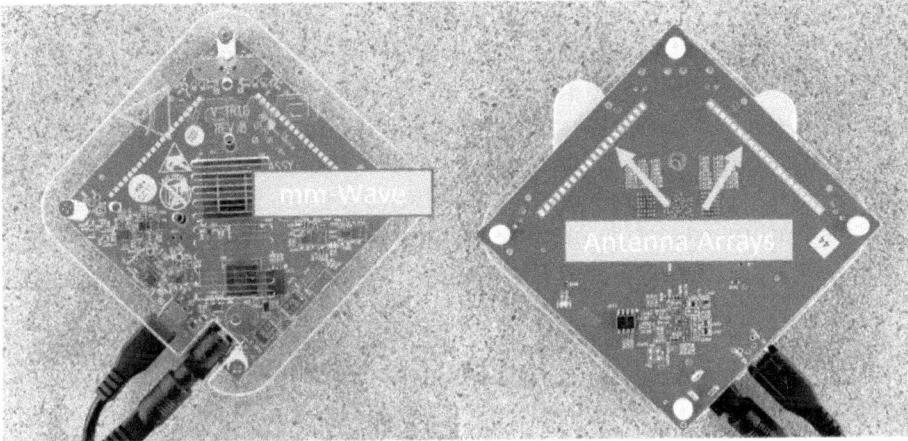

(a) ImageVK-74 mm-waves sensor

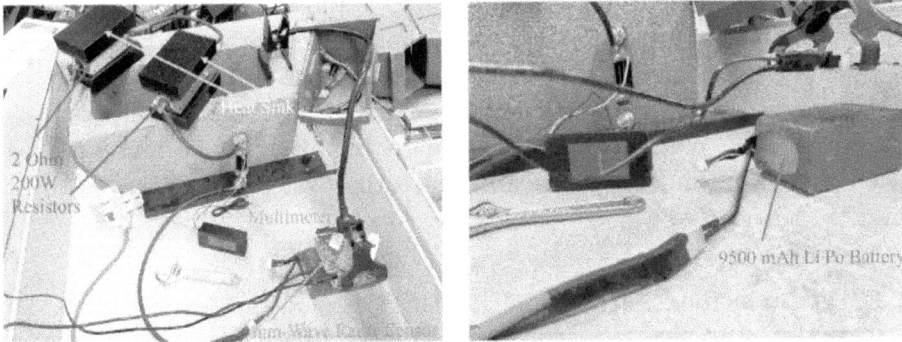

(b) Millimeter wave sensing system for battery

Figure 5.8 ImageVK-74 mm-waves sensor and its millimeter wave sensing system.

simulate the battery used for drone power consumption. The objectives are to test the batteries, including: 1) real battery capacities and, 2) obtaining the real cutoff voltage. The load circuit we deployed is shown in Figure 5.8(b) left. We connect two 2 Ω resistors in series (total external resistance is 4.1 Ω). This load circuit would generate around 5 to 6 A of current for fast discharging. Moreover, the mm-waves sensor is placed above the battery at less than 2 cm height to keep sampling at every 1000 mAh of consumption by monitoring the loading circuit, such as the power cost. During the charging and discharging process of the Li-Po battery, the variation of terminal density during charging and discharging would be captured by mm-waves radar.

5.4.2   Machine learning with mm-waves signals

In this section, machine learning algorithms are selected to process the sample of the mm-wave signals that categorize the battery capacity and voltage. We select principal component analysis (PCA) to reduce high dimensions and use linear support vector machine (SVM) to do classification.

A. PCA algorithm

With data including information about different electrolyte states and LIBs voltages, PCA is always chosen to do the data processing and voltages classifications. Generally, the original data always has multiple information, sampling noise or high dimensions, which is hard to processing and is not necessary. Hence, PCA and linear discriminant analysis (LDA) algorithm are always designed as the reduced-dimension methods for feature extraction of data information processing. It means that certain linear input coordinates are found to obtain a linear projection from the original data space R^n to R^q for some $q < n$.

The aim of PCA is to preserve as much as possible the variance of the original data while reduce the redundancy by transforming to a new set of variables, called principal components. In this section, the built-in PCA module of *scikit learn (sklearn)* is employed to realize the PCA machine learning algorithm. In *sklearn*, it has a module called *sklearn.decomposition.PCA*, which uses full singular value decomposition (SVD) or a randomized truncated SVD of the data to project it to a lower dimensional space. Several typical methods included in the module *sklearn.decomposition.PCA* are listed in Table 5.3. A battery signal model can be constructed with method *fit_transform(X,[y])* and collected signal data by ImageVK-74 mm-waves sensor. Based on this signal model, PCA can reduce the battery data dimension and filter noise of the collected signals.

Table 5.3 Typical methods included in sklearn PCA module

Method Name	Function
fit(X,[y])	Fit the model with X
transform(X,[y])	Apply dimensionality reduction to X
fit_transform(X,[y])	Fit model and transform together
score(X,[y])	Return sample average log-likelihood

B. SVM algorithm

A SVM generates a hyper-plane or group of hyper-planes in a high-dimensional or infinite-dimensional space, which can be used for classification, regression, and other tasks. Intuitively, a good separation is achieved by the hyperplane that is farthest from the training data points of any class. This distance is called the "functional margin", and the higher it is, the smaller the classifier's generalization error. The mathematical formula we used for linear SVM is shown below.

Linear SVM aims to find the maximum-margin hyperplane w and bias term b by minimizing the loss function,

$$\min_{w,b} \frac{||w||^2}{2} + C\sum_{i=1}^{i} max(1 - y_i(w^T\phi(x_i) + b), 0), \tag{5.1}$$

where C denotes the penalty settings used to balance the regularization term and the training loss for labeled and unlabeled data, respectively, SVMs typically discover boundaries in regions with sparsely labeled and imbalanced data.

To solve the multi-class classification problem, N binary SVM classifiers are created in the case of N-class issues ($N > 2$). The ith SVM is trained so that samples in the ith class are labeled as positive and all other samples are labeled as negative. In the classification process, a test sample is obtained from all N SVMs and is labeled based on the classifier with the highest output [3].

In addition, we used the Classification Learner App from Matlab[1] to implement the classification mission. The Classification Learner is used to train classifier models using many popular models such as decision trees, discriminant analysis, support vector machines, logistic regression, closest neighbors, naive Bayes, kernel approximation, ensembles, and neural networks. Moreover, 10-fold cross-validation is used to tweak hyper-parameters and validate the performance of all models on our soil carbon samples. The accuracy and confusion matrices are used to sense how well the objects have been categorized.

5.4.3 Estimation results of battery capacity by mm-wave sensor

The sampling frequency of the mm-wave sensor is from 62 GHz to 69 GHz. We sampled each sample 20 times for the purpose of reducing likelihood error. We set up 150 sampling points during the frequency band we mentioned, and the RBW is 100 kHz for reasonable sweep time and lower noise. As mentioned before, PCA and linear SVM are applied to reduce the data dimension and perform multi-class classification, respectively. The original data scatter plot is shown in Figure 5.9, which has 8 independent classes from 100 mAh to 7100 mAh.

To evaluate our system's performance fairly, we introduced a confusion matrix to calculate average classification accuracy. The 10-fold cross-validation (See Figure 5.10) is also employed to adjust hyper-parameters and validate all models' performance on our battery capacity samples. The accuracy of categorized capacity is 98.8% from the confusion matrix after 10-fold cross-validation. Compared to the classification accuracy (93.75%) of battery voltage by using the Walabot microwave radar in Section 5.3, the accuracy using mm-waves radar is higher.

In addition, we also investigated the relationship between mm-wave signal energy and battery voltage drops in Figure 5.11. As shown in the figure, the mm-wave signal energy decreases slightly as capacity decreases and the voltage drops from 23.5 V to 20.0 V. As we know, the voltage drops of Li-Po batteries are non-linear, and we need to further explore the relationship of mm-wave signal energy with the capacity to predict the battery health, which may help the earlier battery alarm to whether to trigger emergency landing of the drone or not.

In this example of smart sensor for battery, we proposed a contactless method of measuring battery capacity by identifying the reflectance signals of mm-wave radar. The accuracy of measuring capacity is higher than microwave radar. However, there are still some issues that need to be further investigated. First, the sensing distance affects the sensing accuracy. The mm-wave radar sensor will be placed more than 3 or 4 cm above the battery. Second, when testing battery capacity, the battery should not be moved for at least 10 minutes for stable electrolyte density. Finally, the ImageVK-

---

[1]https://www.mathworks.com/help/stats/classificationlearner-app.html

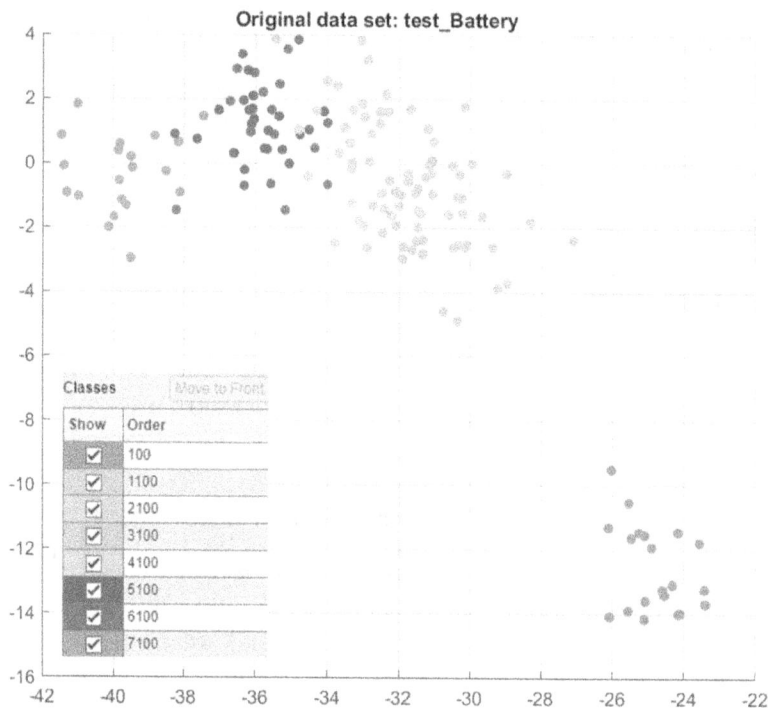

Figure 5.9 Original mm-wave data from varying battery capacity.

74 sensor has a total of 40 antennas for transmitting and receiving signals, referred to as a sensor array. The optimization of number of antennas the sensor array to sense the electrolyte density and Li-ion distribution would be a worthy investigation to both reduce the cost and maintain the reliable battery health awareness.

5.5 CHAPTER SUMMARY

In this chapter, we introduced smart perception for LIBs management to enhance inner knowledge informed by more sensor signals. Mainly five types of sensors are discussed, that is, potential sensor, pressure sensor, thermal sensor, gas sensor, and acoustic sensor. Two examples are presented. One example is a radio-frequency sensor called Walabot for battery voltage classification. This contactless sensor Walabot is combined with CNN algorithm to collect and construct battery images called "battery face". After preprocessing and training with these "battery face" images, CNN algorithm can conduct voltage classification for LIBs. The other example is a millimeter-waves sensor called ImageVK-74, which provides a contactless method of measuring battery capacity by identifying the reflectance signals of this mm-wave radar. The battery capacity can be estimated by the ImageVK-74 sensor and machine learning algorithms (PCA and SVM). The experimental results show that the accuracy of measuring capacity is higher than Walabot microwaves radar.

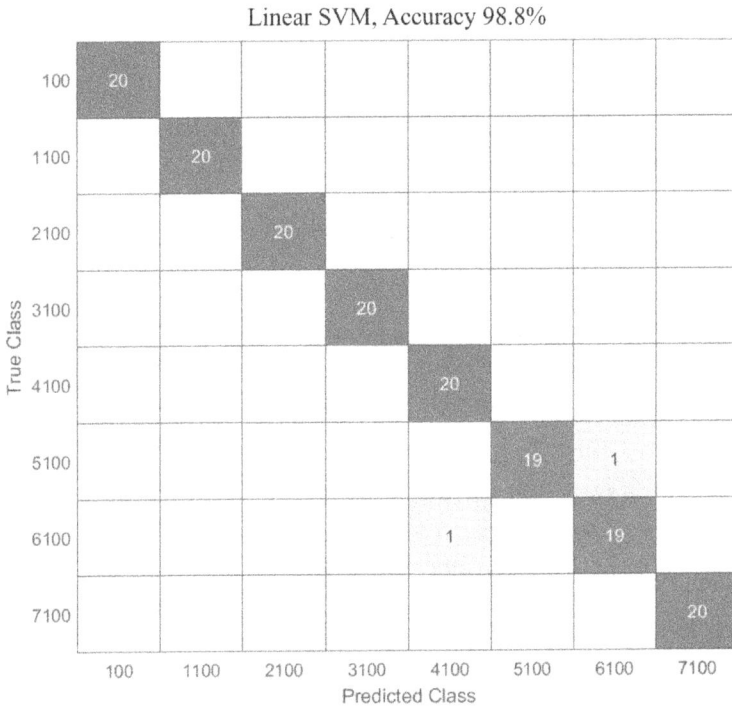

Figure 5.10 Confusion matrix of SOC classification of Li-Po battery.

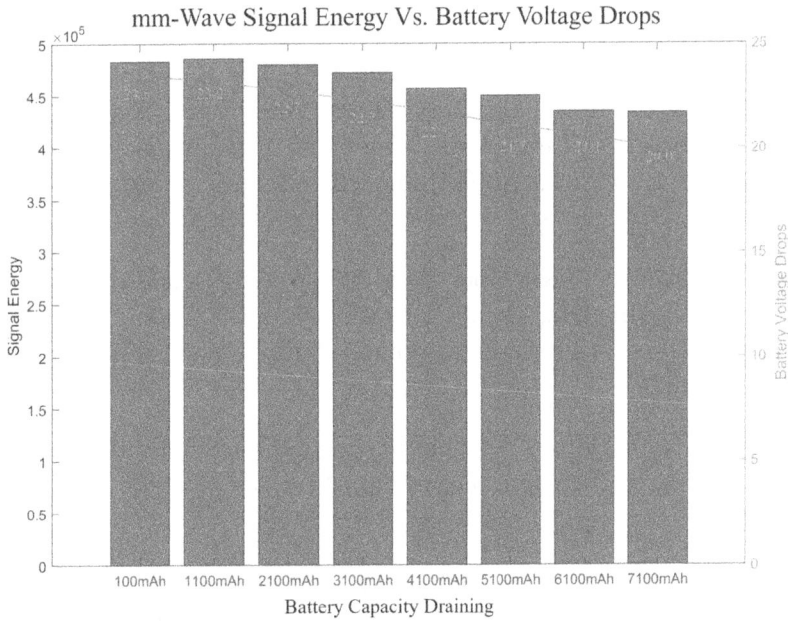

Figure 5.11 Comparison of mm-wave signal energy and battery voltage drops.

The smart sensors applied to battery are not limited to Walabot and ImageVK-74 sensor mentioned in this chapter. It also should be noted that, although more sensors mean more information collected, it also brings more risk and influence for battery operation, which remains to be improved. Moreover, the state-of-art sensors are mostly installed and attached to battery surface rather than inside, thus embedding technology also remains to be improved for further smart perception application.

Bibliography

[1] Laura Albero Blanquer, Florencia Marchini, Jan Roman Seitz, Nour Daher, Fanny Bétermier, Jiaqiang Huang, Charlotte Gervillié, and Jean-Marie Tarascon. Optical sensors for operando stress monitoring in lithium-based batteries containing solid-state or liquid electrolytes. *Nature Communications*, 13(1):1153, 2022.

[2] Di An. *Explainable Artificial Intelligence Internet of Things (XAIoT) Enabled Smart Sensing of Soil Carbon Content for Smart Application of Biochar*. University of California, Merced, 2024.

[3] Fereshteh Falah Chamasemani and Yashwant Prasad Singh. Multi-class support vector machine (SVM) classifiers – an application in hypothyroid detection and classification. In *Proceedings of 2011 Sixth International Conference on Bio-Inspired Computing: Theories and Applications*, pages 351–356, 2011.

[4] Dongying Chen, Qiang Zhao, Yi Zheng, Yuzhe Xu, Yonghua Chen, Jiasheng Ni, and Yong Zhao. Recent progress in lithium-ion battery safety monitoring based on fiber Bragg grating sensors. *Sensors*, 23(12):5609, 2023.

[5] Jiaqiang Huang, Laura Albero Blanquer, Julien Bonefacino, Eric R Logan, Daniel Alves Dalla Corte, Charles Delacourt, Betar M Gallant, Steven T Boles, JR Dahn, Hwa-Yaw Tam, et al. Operando decoding of chemical and thermal events in commercial Na (Li)-ion cells via optical sensors. *Nature Energy*, 5(9):674–683, 2020.

[6] Peixue Jia, Qixiang Zhang, Ziqi Ren, Jianyu Yin, Dandan Lei, Wenzhong Lu, Qianqian Yao, Mingfang Deng, Yihua Gao, and Nishuang Liu. Self-powered flexible battery pressure sensor based on gelatin. *Chemical Engineering Journal*, 479:147586, 2024.

[7] Yann LeCun, Léon Bottou, Yoshua Bengio, and Patrick Haffner. Gradient-based learning applied to document recognition. *Proceedings of the IEEE*, 86(11):2278–2324, 1998.

[8] Sen Li, Yu Cheng, Ka Deng, and Hongyan Sun. A self-powered flexible tactile sensor utilizing chemical battery reactions to detect static and dynamic stimuli. *Nano Energy*, 124:109461, 2024,

[9] Yongkun Li, Chuang Wei, Yumao Sheng, Feipeng Jiao, and Kai Wu. Swelling force in lithium-ion power batteries. *Industrial & Engineering Chemistry Research*, 59(27):12313–12318, 2020.

[10] Kailong Liu, Yuhang Liu, Shiwen Zhao, Xiaoyu Li, and Qiao Peng. An ultrasonic wave-based method for efficient state-of-health estimation of li-ion batteries. *IEEE Transactions on Industrial Electronics*, pages 1–11, 2024.

[11] Oleg Lupan, Nicolai Ababii, Abhishek Kumar Mishra, Ole Gronenberg, Alexander Vahl, Ulrich Schurmann, Viola Duppel, Helge Kruger, Lee Chow, Lorenz Kienle, et al. Single CuO/Cu2O/Cu microwire covered by a nanowire network as a gas sensor for the detection of battery hazards. *ACS Applied Materials & Interfaces*, 12(37):42248–42263, 2020.

[12] Nawei Lyu, Yang Jin, Shan Miao, Rui Xiong, Huaxing Xu, Jinfeng Gao, Haoyan Liu, Yupei Li, and Xiaobei Han. Fault warning and location in battery energy storage systems via venting acoustic signal. *IEEE Journal of Emerging and Selected Topics in Power Electronics*, 11(1):100–108, 2021.

[13] Pariya Nazari, Rainer Báuerle, Johannes Zimmermann, Christian Melzer, Christopher Schwab, Anna Smith, Wolfgang Kowalsky, Jasmin Aghassi-Hagmann, Gerardo Hernandez-Sosa, and Uli Lemmer. Piezoresistive free-standing microfiber strain sensor for high-resolution battery thickness monitoring. *Advanced Materials*, 35(21):2212189, 2023.

[14] Julius Schmitt, Benjamin Kraft, Jan Philipp Schmidt, Betina Meir, Klaus Elian, David Ensling, Goran Keser, and Andreas Jossen. Measurement of gas pressure inside large-format prismatic lithium-ion cells during operation and cycle aging. *Journal of Power Sources*, 478:228661, 2020.

[15] Yi Shen, Sheng Wang, Haomiao Li, Kangli Wang, and Kai Jiang. An overview on in situ/operando battery sensing methodology through thermal and stress measurements. *Journal of Energy Storage*, 64:107164, 2023.

[16] Ting Tang, Quan Xia, Mingkang Xu, Zhe Deng, Fusheng Jiang, Zeyu Wu, Yi Ren, Dezhen Yang, and Cheng Qian. Uneven internal soc distribution estimation of lithium-ion batteries using ultrasonic transmission signals: A new data screening technique and an improved deep residual network. *eTransportation*, page 100406, 2025.

[17] Weihan Wang, Yanyun Zhang, Bin Xie, Lang Huang, Shanmu Dong, Gaojie Xu, and Guanglei Cui. Deciphering advanced sensors for life and safety monitoring of lithium batteries. *Advanced Energy Materials*, 14(24):2304173, 2024.

[18] Xudong Xia, Wen Wu, Zhencheng Li, Xile Han, Xiaobin Xue, Gaozhi Xiao, and Tuan Guo. State of charge estimation for commercial Li-ion battery based on simultaneously strain and temperature monitoring over optical fiber sensors. *IEEE Transactions on Instrumentation and Measurement*, 73:1–11, 2024.

[19] Song Xie, Zhipeng Wang, Ju Fu, Pengfei Lv, and Yuanhua He. A review of sensing technology for monitoring the key thermal safety characteristic parameters of lithium-ion batteries. *Journal of Power Sources*, 624:235598, 2024.

[20] Yi Zhang, Xiangpeng Xiao, Weilun Chen, Zihan Zhang, Wanming Li, Xiaoyu Ge, Yanpeng Li, Jingwei Xiang, Qizhen Sun, Zhijun Yan, et al. In operando monitoring the stress evolution of silicon anode electrodes during battery operation via optical fiber sensors. *Small*, 20(29):2311299, 2024.

[21] Zhiguo Zhang, Yiding Li, Xueqing Min, Dongsheng Ren, Youzhi Song, Li Wang, Hong Zhao, and Xiangming He. Enhancing precision and durability of built-in Cu-Li reference electrodes in lithium-ion batteries: A critical review. *ACS Energy Letters*, 9(11):5647–5669, 2024.

Perspectives on Intelligent Fractional-Order Modeling

This chapter offers an overview of intelligent fractional-order modeling for batteries during their entire lifetime. We highlight the reasons to revisit modeling in general and fractional-order modeling in particular for batteries, and try to provide an integrated perspective on the intelligent management system for the next generation BMS. Demands, opportunities, and challenges are presented in the following sections.

6.1   INTELLIGENT MANAGEMENT IN BATTERY LIFESPAN

With the development of artificial intelligence (AI), intelligence for battery is designed from different working stages and multi-dimensions. A smart battery with intelligent management system could be achieved as shown in Figure 6.1, and the life of a battery is generally divided into six stages as design, manufacturing, monitoring, control, protection, and recycling (numbered as layer1–layer6). The smart battery system in Figure 6.1 includes design and manufacture stage (layer1–layer2), because battery issues may inherently exist before battery usage, and intelligent management is necessary from the beginning to solve battery inner problem. Among the six stages, control and protection would work in parallel during vehicle operation. Recycling is also considered for further management (such as cascade utilization and material recycle) when battery retirement. All six layers are embedded with AI to form a "smart" inter-connected entity during the whole battery lifetime. With this "smart" network, various kinds of battery data and information can be collected from the six layers and gathered together in the cloud, which is called AI management in the top layer of Figure 6.1. Besides, as shown from the bottom to the top of Figure 6.1, the six layers of smart battery system transfers from material science (layer1–layer2, battery design and manufacture) to systematic science (layer3–layer6), then comes back to material science in the end of layer6 (recycling to original material).

During the six stages shown in Figure 6.1, for various operation temperatures and the whole life working of battery, several issues remain unsolved, such as health evaluation, battery sorting, and knee point prediction. The core problem of these issues is about the accurate and reliable battery state information. Traditional modeling

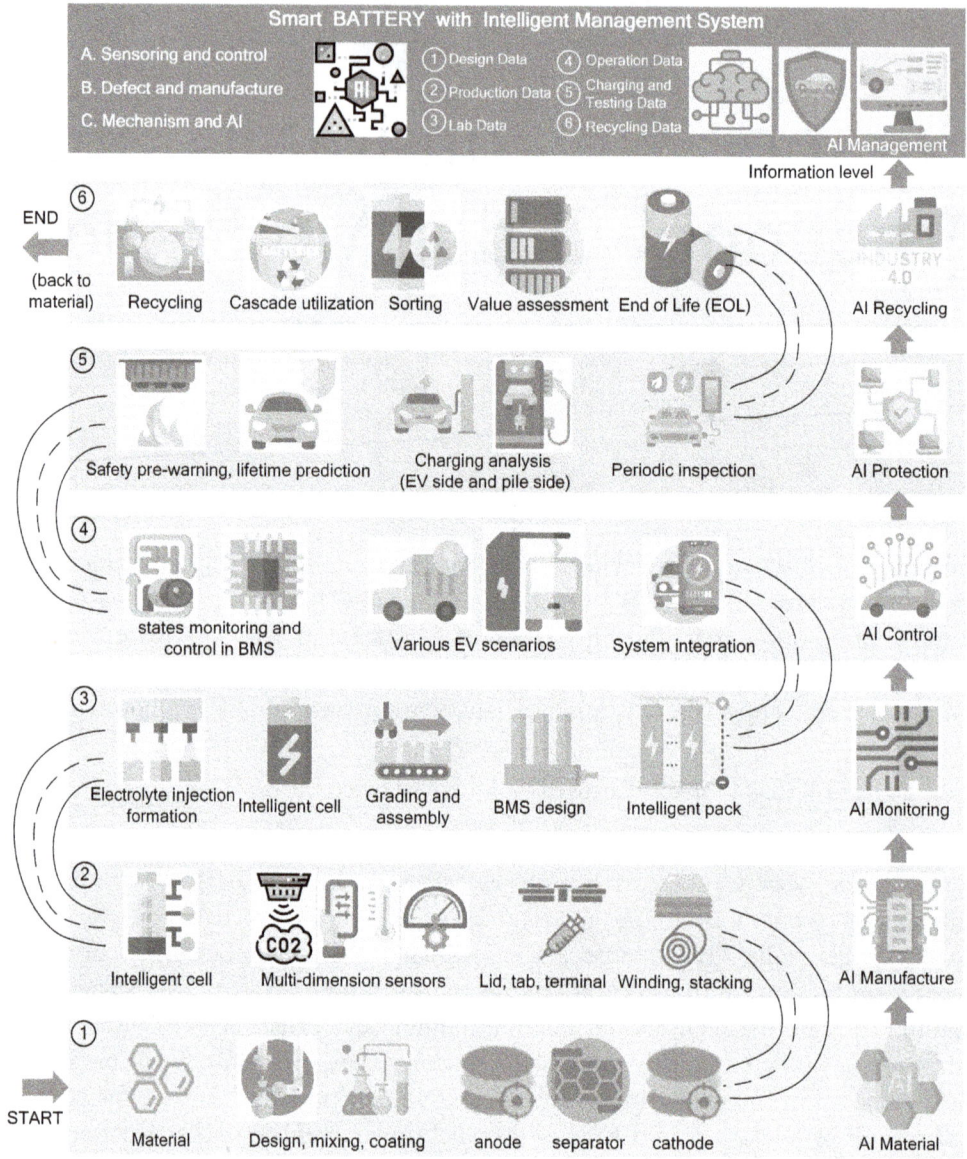

Figure 6.1 The overview of smart battery with intelligent management system. AI in the figure stands for intelligence, smart design, and wise strategy in every aspect and layers. The framework contains six layers for six battery stages (numbered from 1 to 6), which are intelligent material design (AI+Material), intelligent manufacturing (AI+Manufacture), intelligent cell with intelligent BMS (AI+Monitoring), intelligent system control (AI+Control), intelligent safety and lifetime protection (AI+Protection), intelligent sorting and cascade utilization (AI+Recycling), and big-data algorithm (AI+Management). The smart battery covers from material level (layer 1, layer 2) to system level (layer3–layer6), and collects data in every layer to construct an information level in the cloud (called AI+Management). The six layers form a cycle starting from material and back to material by battery recycling in the end.

technologies, such as electrochemical model and ECM, can obtain the battery state accurately only on several conditions. While the application of big data and AI technologies embedded with battery mechanism knowledge still cannot maintain stable results for battery working in various operation temperatures and the whole lifetime. Moreover, based on accurate state estimation, real-time control strategy remains to be further developed, to achieve a closed-loop feedback for the intelligent management system.

6.2   INTELLIGENT MODELING FOR BATTERY

With the forward-looking overview shown in Figure 6.1, battery with intelligent management during life cycle relates to multidisciplinary fields from material science to system science. Problem-oriented investigations, demands and foundational technology are presented in this section and then modeling trends are provided to offer several possibilities in the next generation BMS.

6.2.1   Demands

From Figure 6.1, intelligent management with six stages in battery lifespan require multidisciplinary knowledge from material level to system level, including chemistry, material, thermal engineering, electrical engineering, mechanical engineering, and control [1]. The demand of battery modeling with AI focuses on the improvement of intelligence technology and acquisition of lifetime information, and these improvements are mainly in AI control, AI Management, and AI Recycling levels. Battery working in various temperatures and lifecycle (including recycling stage) mainly requires four demands, that is, AI algorithm iteration, battery mechanism supervision, big-data platform support, and multi-scenario extension [22], as follows.

- AI algorithm iteration can introduce powerful machine learning (ML) algorithm for intelligent BMS.

- Battery mechanism for battery knowledge embedding can bring physical meanings to AI algorithm.

- Big-data platform needs to be designed for standardized data processing, key information collection, and efficient usage.

- Multi-scenario extension is needed to achieve foundation model, which can be used in different vehicles, batteries, operation conditions, and life stages.

6.2.2   Foundational technology

For the existing issues and demand through the six stages of battery lifespan in Figure 6.1, every stage in the framework may include specific tasks and technologies to be developed, three foundational technologies are suggested here, as shown in Figure 6.2.

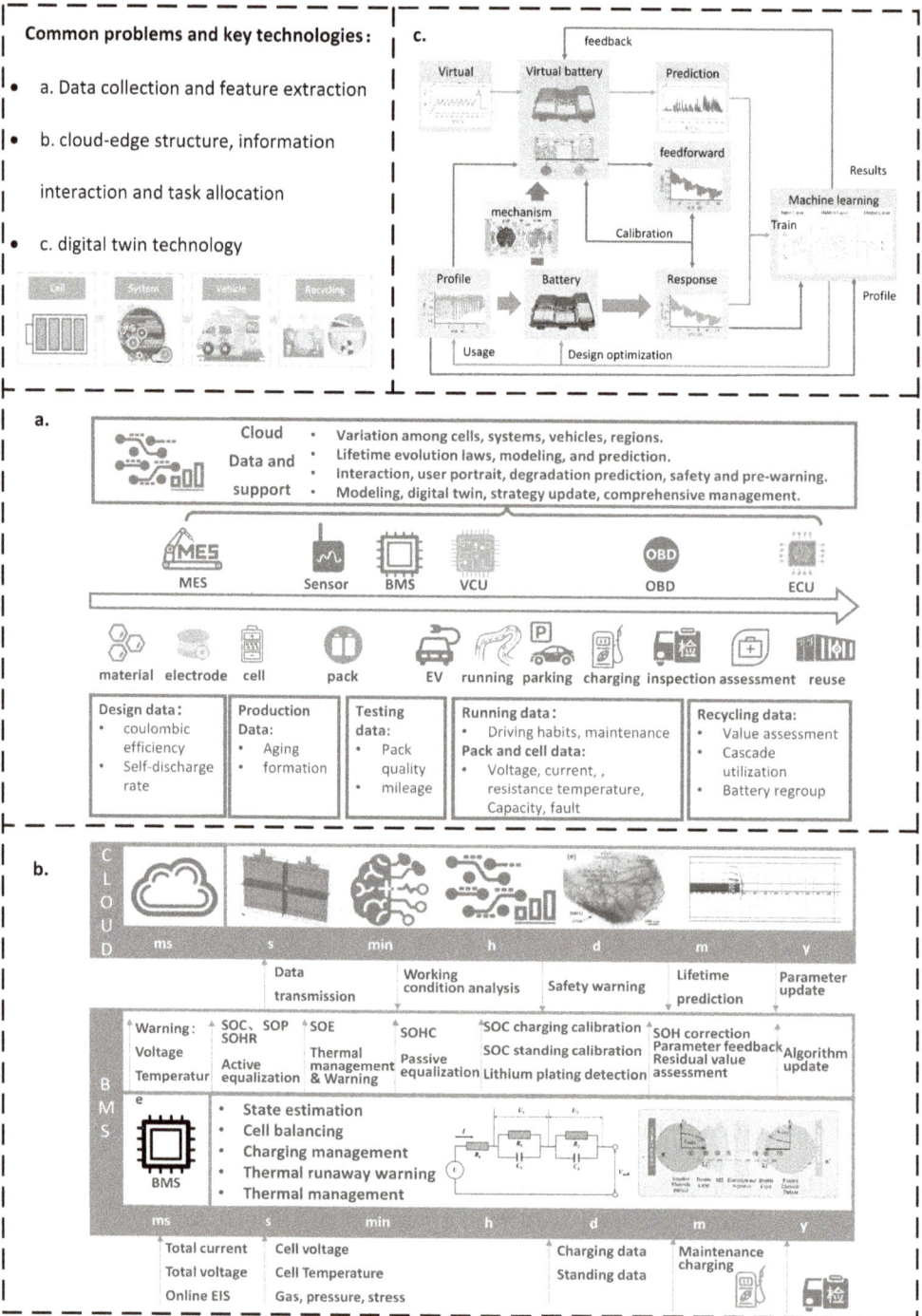

Figure 6.2 Three foundational technologies are also presented in this framework as a. distribution of flow field in lifespan, (b) cloud-edge-end structure, and (c) DT and data augmentation.

- Distribution of flow field in battery lifespan: as shown in Figure 6.2(a), material in battery goes through original material, electrode, cell, pack, vehicle, operation (running, parking, charging, and inspection), assessment, reuse, recycle, finally back to material. There is a material and energy flow through the whole battery lifespan. In the material and energy flow, data can be collected at different levels from design, production, testing, running (pack and cell), to recycle level. All collected data can be further sent to the cloud center. With the data collection, a data information flow through the whole battery lifespan is also built. With knowing the distribution of the closed-loop flow field, the system can make full use of the energy exchange and data information, then lifespan monitoring and battery portrait can be actually achieved.

- Cloud-edge-end structure : BMS in the next generation can be designed with the development of AI, cloud computing, edge computing, Internet of Thing (IoT), and 5G technology [21, 23]. The connection between physical world and virtual world can be built by end device, edge device, and cloud platform, as shown in Figure 6.2(b). End device may include sensors, embedded BMS, and production line, which responsible for data quality and privacy by data collection, preprocess, characteristic extraction, and basic monitoring functions. Edge mainly means charging pile and inspection rather than the "edge computing" in the AI field, and it is responsible for assisting and providing charging data and specific data under inspection conditions. Cloud is usually a platform with big database and powerful processor for comprehensive calculation, multidimensional modeling, and global analysis.

- Digital twin (DT) and data augmentation: a virtual mapping model with high accuracy can be constructed for objects in reality, resulting in "virtual battery" [20], as shown in Figure 6.2(c). DT technology avoids large amount of experiments and long development period [8], because DT technology may generate data that may not be obtained in reality, such as battery defects and battery incidents. Moreover, data augmentation technology can also have similar effects and be deployed to generate virtual data under dynamical operation circumstances. Combined with cloud-edge-end structure, DT technology can be deployed to achieve a mechanical-electro-thermal mechanism model, which is hardly used in traditional BMS [13].

6.2.3 Modeling trends

To illustrate the modeling trends, we take battery safety as the focus in this section and the modeling methods can be divided into three main types as shown in Figure 6.3. On the basis of battery mechanism analysis, states monitoring and management can be improved with algorithm development [7, 27]. First of all, data source is always the primary step for states monitoring. Big data during operation should be carefully collected in the cloud platform, then feature dataset can be deduced from these big data for further algorithm development. Besides data collection, the development of battery intelligent modeling can be mainly divided into three types, that is, mechanism-based method (Figure 6.3 A), data-driven method (Figure 6.3 B), and method with multi-sensors (Figure 6.3 C).

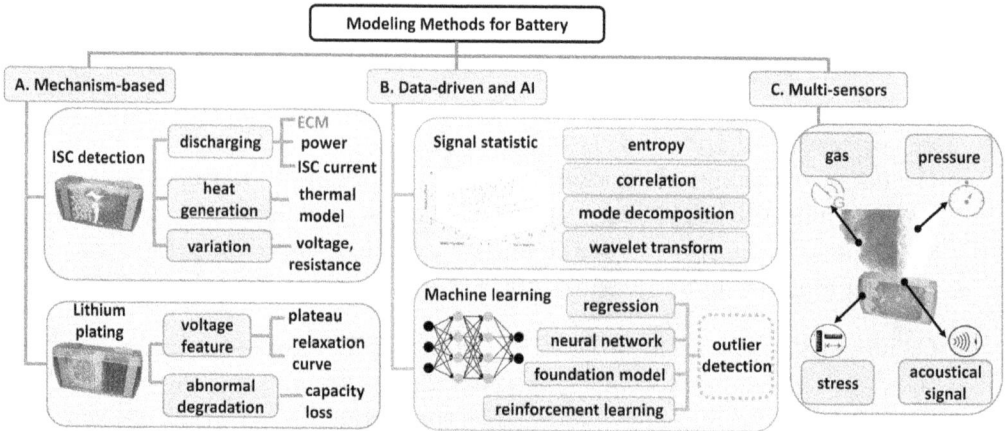

Figure 6.3 Battery modeling for safety monitoring, which can mainly be divided into three classes, that is, A. mechanism-based methods, B. data-driven and AI methods, C. multi-sensors methods.

- Mechanism-based method: internal short circuit (ISC) detection methods such as ECM [3], thermal model [28, 4], and voltage or resistance variation model [15].

- Data-driven method: signal statistic methods such as entropy, correlation, and wavelet transform, and machine learning methods such as regression, neural network, foundation model, and reinforcement learning [24, 26].

- Method with multi-sensors: application of sensor sampling, such as gas, pressure, stress, and acoustical signals [14, 25].

With the three main modeling methods introduced above, integrated modeling can also be achieved [9, 2], such as data-driven method for parameter identification of mechanism model, mechanism information embedded in data-driven method, and sensor signal as input of mechanism model or data-driven model. As shown in Figure 6.3, the three main trends of modeling methods for battery safety (mechanism-based modeling, data-driven modeling, and multi-sensors modeling) are also suitable for other aims of battery management, such as state estimation, lifetime prediction, or health evaluation.

6.3   SMART BATTERY WITH FRACTIONAL-ORDER MODELING

Upon the intelligent management and intelligent modeling for battery, a smart battery system can be constructed. Based on the non-linear characteristics of battery, fractional calculus is always introduced for battery modeling, resulting in smart battery with fractional-order modeling. The core framework for smart battery system in next generation is first proposed, then fractional-order modeling inside this framework is discussed in details to highlight the valuable functions of fractional calculus for smart battery system.

6.3.1 Smart battery system in next generation

From the management aspect, a core framework for battery intelligent management in the next generation BMS is proposed as shown in Figure 6.4. In the center of the management core is the battery life cycle from material to system, finally back to material level, which is similar to the structure shown in Figure 6.1. However, Figure 6.4 focuses on only the management aspect, which also includes six main parts: (1) mechanism, (2) combination, (3) pure AI, (4) cloud software, (5) edge hardware, and (6) retirement well. We will present every part in the following as four sections, that is, A. battery mechanism (1 mechanism), B. AI algorithm (2 combination and 3 pure AI), C. platform support (4 cloud software and 5 edge hardware), finally D. recycling and cascade utilization (6 retirement well).

A. Battery mechanism (1 mechanism)

Battery mechanism acts as the first principle and vital role in the proposed core framework as shown in Figure 6.4. For reinforcement learning or other kind of deep learning, the realization of these algorithms is to realize a "learn to learn" skill, which means to learn the object rules or mechanism. Battery mechanism can act as the instruction for algorithm optimization and control strategy design, resulting in a combination of AI and mechanism. To enhance the interpretability, one kind of method is applying battery mechanism to feature extraction and algorithm structure improvement. If feature engineering and feature extraction are under the instruction of battery mechanism, the modeling algorithm knows which kinds of battery features are worthy of learning. The algorithm interpretability comes from battery mechanism; thus mechanism is the basis and first investigated part in the core framework.

B. AI algorithm (2 combination and 3 pure AI)

As to the development of AI algorithm, pure AI and combination are two main aspects to be concerned for battery life-cycle management. The battery combination models with mechanism and AI can be divided into three types: physics-based model, data-driven model, and hybrid model. Utilization of battery mechanism constructs physics-based models, while utilization of pure AI constructs data-driven models. As shown in Figure 6.4, the combination of AI and mechanism would construct a kind of hybrid model, called interpretable AI. The algorithm interpretability comes from battery mechanism, but the algorithm still holds a basic framework from AI technologies. Structure improvement achieves a kind of interpretable algorithm called physics-informed machine learning (PIML), in which physical laws of battery mechanism can be embedded into the training process, such as forward propagation, backpropagation, and loss calculation. In this way, the algorithm is informed by certain physic information of battery mechanism to instruct the training process, which makes algorithm learn the rule of mechanism. Another kind of interpretable AI is digital twin (DT) or virtual reality (VR). With powerful hashrate and storage ability of cloud platform, DT combines traditional modeling and big data to construct a virtual

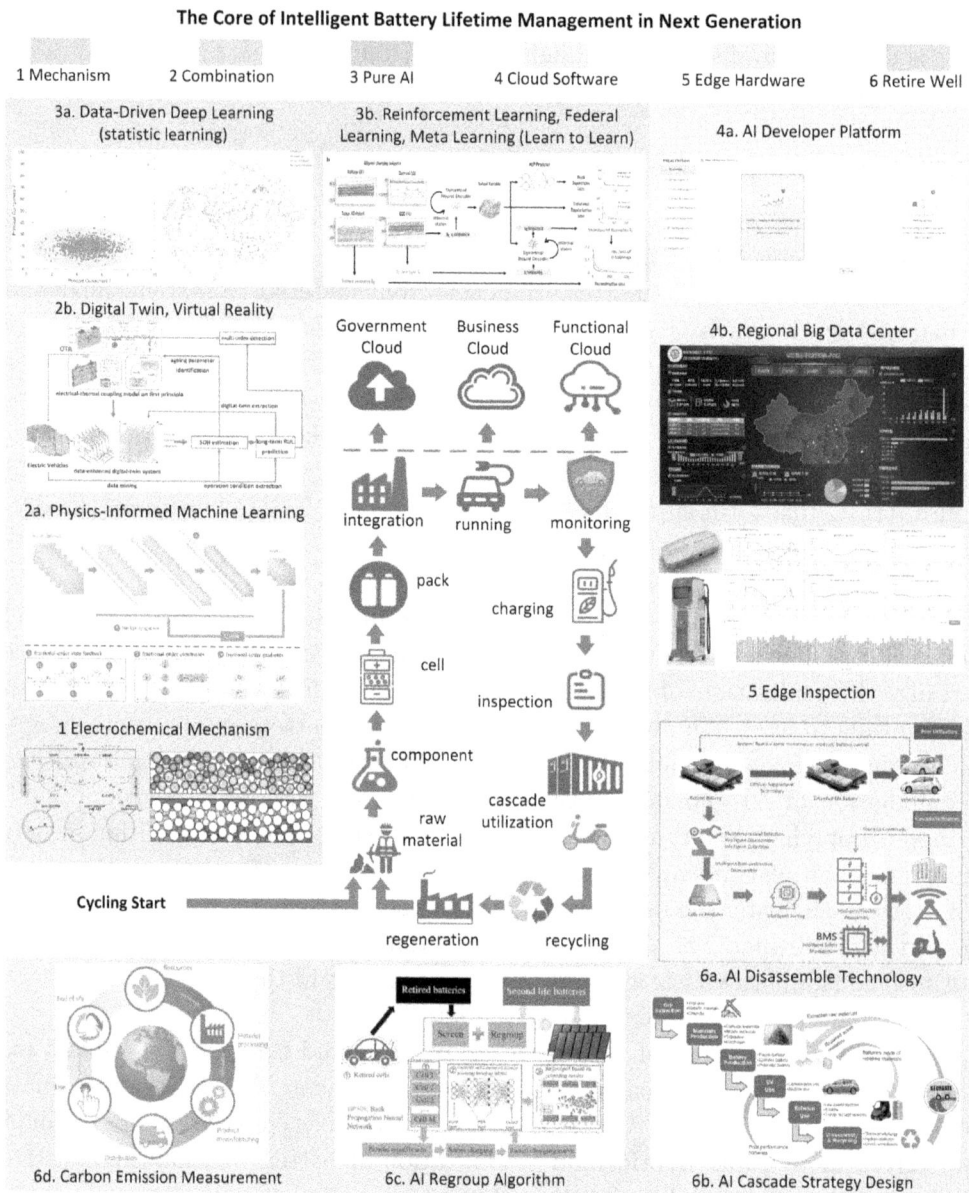

Figure 6.4 The core of smart battery life-cycle management in next generation. Six main parts during battery lifetime cycling: 1. Mechanism, inside electrochemical reactions are the basis and first principle for battery management all the time; 2. Combination of mechanism and data-driven algorithm, such as physics-informed machine learning, DT, and virtual reality; 3. Pure AI, only data-driven algorithm from statistic direction, such as deep learning, reinforcement learning, federated learning, and meta learning; 4. Cloud software, mainly focus on various database and platform construction, such as developer platform, and big-data center; 5. Edge hardware, hardware to support data collection; 6. Retire well, this part includes four key parts to achieve battery well-retirement, that is, disassemble technology, cascade strategy, regroup algorithm, and carbon emission measurement.

data-driven model with high accuracy for battery. Hence, this kind of virtual model build a connection between virtual world ("virtual battery") and reality world (the real battery substance).

Algorithm can also be upgraded by the thriving AI technologies, such as transformer, self-supervised learning, and transfer learning, resulting in pure AI, as shown in Figure 6.4. In recent years, foundation model and its pre-training gradually become mainstream in artificial general intelligence (AGI) field, especially Natural Language Processing (NLP). The Generative AI used in foundation model has been introduced into many other data-driven application fields, such as biotechnology, autopilot, material design, and energy storage. With the new AI era started by ChatGPT, more AI technologies in this direction have drawn much attention in the battery management field.

For algorithm upgrade, the core learning pattern of AI algorithm has transferred from dataset learning to statistical learning to learning-rules learning. Then, simple data-driven deep learning gradually changes to learn-to-learn strategies, such as transfer learning, reinforcement learning, federated learning, and meta learning. However, battery is an electrochemical object with non-linear dynamics, thus battery data always have no labels and unstable quality. In this way, the application of foundation model and AI algorithms may face difficulties such as unsupervised learning, transfer requirement and generality for various types of batteries.

C. Platform support (4 cloud software and 5 edge hardware)

The discussion about AI technologies mentioned in the previous part B is mainly from the aspect of software. Besides, the hardware is another part to realize the intelligent battery life-cycle management in practical. Platform support includes cloud platform and edge hardware. As shown in "4 Cloud Software" of Figure 6.4, cloud platform includes developer platform (for algorithm development) and big-data center (for regional deployment of cloud system). Except for data storage, the cloud platform part is the base of all the algorithms and digital models generated from the "2 Combination" and "3 Pure AI" parts. Edge hardware includes all edge device to conduct edge inspections (safety and health check) as shown in "5 Edge Inspection" (Figure 6.4). The edge inspection part acts as an assistant to support running monitoring and cloud analysis for EVs.

D. Recycling and cascade utilization

The rest of Figure 6.4 is about an important and also the final stage of vehicle battery, called "6 Retirement well" in the core of life-cycle management. Battery retired from EVs may hold potential usage in other working conditions, such as storage system, home appliances, and so on. Hence, with reasonable cascade utilization, retired battery can have a second life and extend its lifespan value. The cascade utilization and recycling of retired battery mainly include four parts: intelligent cascade strategy design, intelligent disassemble technology, AI regroup algorithm, and carbon emission measurement.

6.3.2 Fractional calculus for battery

According to the smart battery life-cycle management shown in Figure 6.4, fractional calculus can enhance the effectiveness of intelligent modeling for battery. The three potential improvements can be concluded as follows as shown in Figure 6.5.

(1) Fractional-order element (CPE) for battery modeling.

The application of CPE is widely used for battery modeling [5], especially under complex working conditions [10, 11]. However, further improvement of

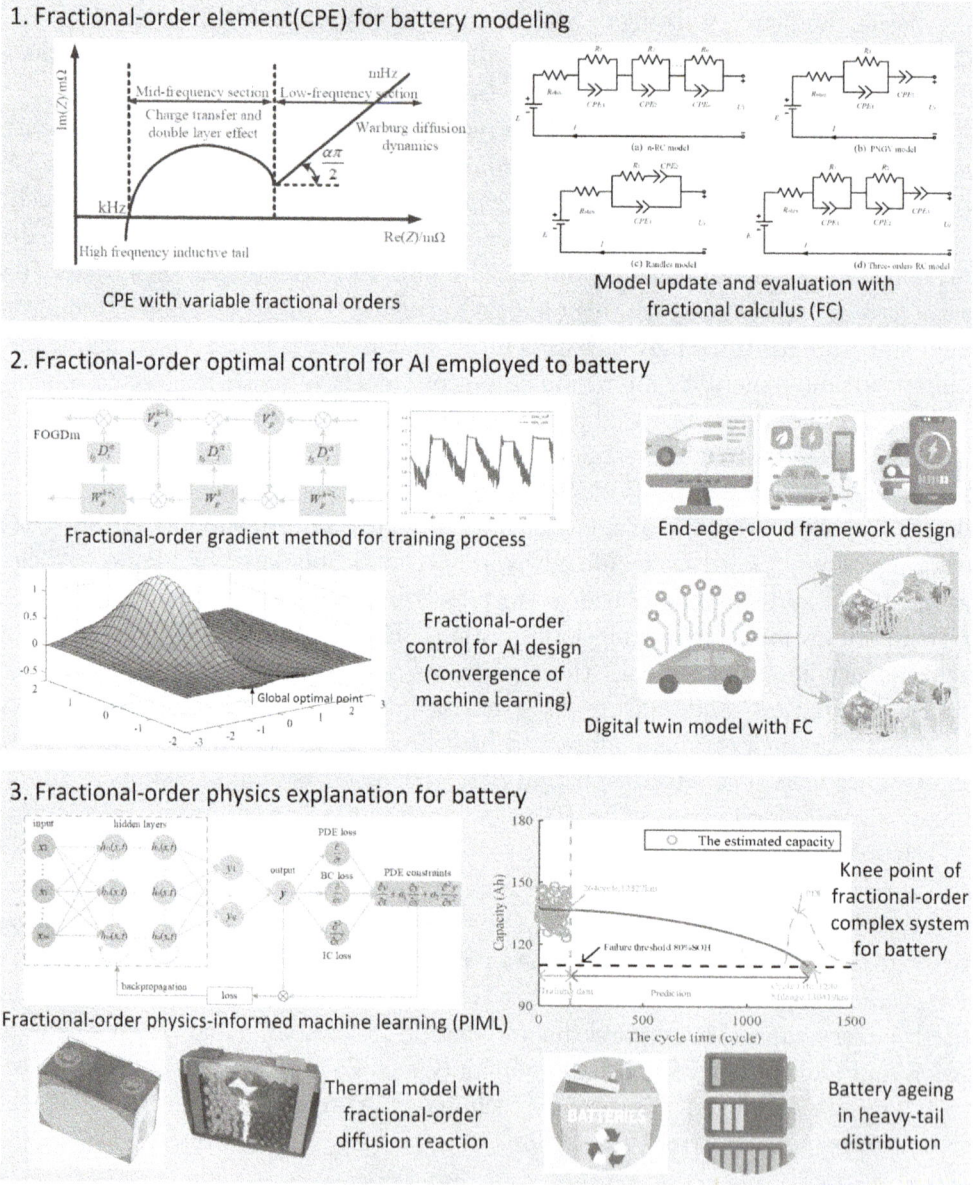

Figure 6.5 Fractional calculus (FC) for battery in three potential aspects: 1. fractional-order element for battery modeling; 2. fractional-order optimal control for AI empowered battery; 3. fractional-order physics explanation for battery.

fractional-order element is worthy of investigation. For example, as shown in the first part of Figure 6.5, the fractional order in CPE can be a variable order instead of a fixed order under different battery states. The parameter identification of battery models can also introduce fractional-order laws (such as inverse power law) for modeling updates and iteration during battery life cycle. Moreover, the effects of battery model need to be evaluated and enhanced, and the probability distribution function (PDF) of model accuracy may be analyzed with variable fractional order and model structure.

(2) Fractional-order optimal control for AI employed in battery.

For AI, machine learning, and deep learning for battery intelligent modeling, fractional calculus can also play a vital role in the optimization of algorithms and control feedbacks [12, 6]. How can researchers determine the optimal AI structure or parameters for battery intelligent modeling? Hence, the AI algorithm needs to be evaluated and optimized, which fractional-order optimal convergence would provide instructions for AI design. For example, as shown in the second part of Figure 6.5, fractional-order gradient methods with backpropagation can be developed for training process. The convergence of machine learning can also be transformed into an optimization problem, which can be resolved by fractional-order control theory. In the systematic aspect, fractional-order optimization can also provide instructions for the design of DT model, edge-AI computing, and end-edge-cloud framework.

(3) Fractional-order physics explanation for battery.

Another thriving aspect is the fractional-order information of battery characteristics embedded into intelligent modeling. As shown in the third part of Figure 6.5, the fractional-order physics-informed machine learning (PIML) presented in this book is a kind of fractional-order explanation for battery [19, 18, 17, 16]. Furthermore, thermal modeling of battery can also be improved by adding fractional-order elements, because thermal diffusion behavior of battery essentially shows fractional-order characteristics. The stochastic and dynamic changes of battery capacity during lifespan degradation may follow a heavy-tailed distribution law or Student's t-distribution phenomenon, which can be modeled as fractional-order model with heavy-tailed noise. Moreover, the knee points of battery degradation can also be predicted as the tipping points for fractional-order complex system.

6.3.3 Opportunities and challenges

Intelligent battery management takes the advantages of safety prognostics or pre-warning and DT at mechanism level and life-cycle management at AI level. Safety pre-warning still needs to clarify the mechanism of battery faults, such as internal short circuit, and morphology of lithium plating, then conduct fault diagnosis and warning. Safety pre-warning may be realized only by data-driven methods, but fault mechanism is still the key for accurate warning for various batteries. DT also needs to apply battery mechanism rather than only data-driven algorithms for optimal performance simulation, state estimation, and evaluation of batteries. Life-cycle management in cloud level may develop an intelligent system with AI technologies for health evaluation, RUL prediction, and degradation analysis of massive battery systems.

Although many opportunities exist for the smart battery with intelligent management system, challenges and risks also accompany all the way to the next generation systems. For example, for the information and energy flow through battery lifespan, the material and data in six layers shown in Figure 6.1 are independent and belong to different manufacturers currently. Hence, a third-party technology should be developed to aggregate all material/energy and data/information in a management platform with database. Most recently, federated learning is becoming an emerging technique for BMS. The intersection between federated learning and battery management systems, focusing on improving battery performance and longevity while ensuring data privacy, may provide a spacious playground for new research and technology development. For example, federated digital twins could be developed as a foundational digital twin, using federated learning with diverse data sets from various sources ranging from household users to EV manufacturers when data privacy preservation is a concern.

Moreover, safety, power, and durability are still the three main aspects concerned for BMS. For safety management, safety modeling, feature extraction, and strategy development are the three key points. For power management, multi-sensors, integration of mechanism with AI, and algorithm development platform are three key breakthroughs in the next generation BMS. For durability, mechanism, management under V2G (vehicle-to-grid) condition, and lifetime evaluation with recycling are the three key points. Integration of battery mechanism and AI is still in the early stage of investigation, and the efficiency of AI technologies may be inconsistent for different batteries or datasets.

6.4 CHAPTER SUMMARY

In this chapter, we presented several perspectives on potential future development opportunities for BMS and intelligent fractional-order modeling. Smart battery with intelligent management is proposed as a new framework for future work with six levels defined during the entire battery life-cycle. In this framework, the future requirements or demands and foundational technologies such as digital twins are analyzed for intelligent modeling for battery in the next generation. Development trends are also discussed in the three main model types, that is, mechanism-based model, data-driven model, and model with multi-sensors. Smart battery with fractional-order modeling is highlighted in this chapter to introduce several potential improvements by fractional calculus in three aspects. They are fractional calculus, fractional-order elements, fractional-order control that can bring new view for battery modeling, parameter identification, structure design, algorithm convergence, AI development, and even end-edge-cloud framework optimization. We attempt to suggest some new ideas about applying fractional-order information into battery and its intelligent modeling in this chapter for further research.

Bibliography

[1] Mohammad Alkhedher, Aghyad B Al Tahhan, Jawad Yousaf, Mohammed Ghazal, Reza Shahbazian-Yassar, and Mohamad Ramadan. Electrochemical and thermal modeling of lithium-ion batteries: A review of coupled approaches for improved thermal performance and safety lithium-ion batteries. *Journal of Energy Storage*, 86:111172, 2024.

[2] Mahshid N Amiri, Anne Haakansson, Odne S Burheim, and Jacob J Lamb. Lithium-ion battery digitalization: Combining physics-based models and machine learning. *Renewable and Sustainable Energy Reviews*, 200:114577, 2024.

[3] Zhoujian An, Tianlu Shi, Xiaoze Du, Xian An, Dong Zhang, and Jianhua Bai. Experimental study on the internal short circuit and failure mechanism of lithium-ion batteries under mechanical abuse conditions. *Journal of Energy Storage*, 89:111819, 2024.

[4] Santosh Chavan, Bhumarapu Venkateswarlu, Mohammad Salman, Jie Liu, Prakash Pawar, Sang Woo Joo, Gyu Sang Choi, and Sung Chul Kim. Thermal management strategies for lithium-ion batteries in electric vehicles: Fundamentals, recent advances, thermal models, and cooling techniques. *International Journal of Heat and Mass Transfer*, 232:125918, 2024.

[5] Saddam Gharab, Asma Achnib, Patrick Lanusse, and Vicente Feliu Batlle. Fractional order modeling of lithium-ion batteries for a real smart grid system. *IFAC-PapersOnLine*, 58(12):478–483, 2024.

[6] Wessam A Hafez, Mokhtar Aly, Emad A Mohamed, and Nadia A Nagem. Improved fractional order control with virtual inertia provision methodology for electric vehicle batteries in modern multi-microgrid energy systems. *Journal of Energy Storage*, 106:114582, 2025.

[7] Xin Lai, Wei Yi, Xiangdong Kong, Xuebing Han, Long Zhou, Tao Sun, and Yuejiu Zheng. Online detection of early stage internal short circuits in series-connected lithium-ion battery packs based on state-of-charge correlation. *Journal of Energy Storage*, 30:101514, 2020.

[8] Weihan Li, Monika Rentemeister, Julia Badeda, Dominik Jóst, Dominik Schulte, and Dirk Uwe Sauer. Digital twin for battery systems: Cloud battery management system with online state-of-charge and state-of-health estimation. *Journal of Energy Storage*, 30:101557, 2020.

[9] Jie Liu, Saurabh Yadav, Mohammad Salman, Santosh Chavan, and Sung Chul Kim. Review of thermal coupled battery models and parameter identification for lithium-ion battery heat generation in EV battery thermal management system. *International Journal of Heat and Mass Transfer*, 218:124748, 2024.

[10] Wei Liu, Jiashen Teh, and Jian Shi. A review of lithium-ion battery models. *Battery-Integrated Residential Energy Systems*, pages 33–59, 2024.

[11] Rohit Mehta and Amit Gupta. Mathematical modelling of electrochemical, thermal and degradation processes in lithium-ion cells—a comprehensive review. *Renewable and Sustainable Energy Reviews*, 192:114264, 2024.

[12] K Dhananjay Rao, Anilkumar Chappa, SVNSK Chaitanya, A Hemachander, B Phani Teja, Subhojit Dawn, Miska Prasad, and Taha Selim Ustun. Fractional order modeling based optimal multistage constant current charging strategy for lithium iron phosphate batteries. *Energy Storage*, 6(2):e593, 2024.

[13] Concetta Semeraro, Haya Aljaghoub, Mohammad Ali Abdelkareem, Abdul Hai Alami, Michele Dassisti, and AG Olabi. Guidelines for designing a digital twin for Li-ion battery: A reference methodology. *Energy*, 284:128699, 2023.

[14] Dongxu Shen, Dazhi Yang, Chao Lyu, Jingyan Ma, Gareth Hinds, Qingmin Sun, Limei Du, and Lixin Wang. Multi-sensor multi-mode fault diagnosis for lithium-ion battery packs with time series and discriminative features. *Energy*, 290:130151, 2024.

[15] Jiaqiang Tian, Yuan Fan, Tianhong Pan, Xu Zhang, Jianning Yin, and Qingping Zhang. A critical review on inconsistency mechanism, evaluation methods and improvement measures for lithium-ion battery energy storage systems. *Renewable and Sustainable Energy Reviews*, 189:113978, 2024.

[16] Yanan Wang, Xuebing Han, Feng Dai, Jie Li, Daijiang Zou, Languang Lu, Yangquan Chen, and Minggao Ouyang. Fractional order backpropagation neural network for battery capacity estimation with realistic vehicle data. In *2022 18th IEEE/ASME International Conference on Mechatronic and Embedded Systems and Applications (MESA)*, pages 1–6. IEEE, 2022.

[17] Yanan Wang, Xuebing Han, Dongxu Guo, Languang Lu, Yangquan Chen, and Minggao Ouyang. Physics-informed recurrent neural network with fractional-order gradients for state-of-charge estimation of lithium-ion battery. *IEEE Journal of Radio Frequency Identification*, 6:968–971, 2022.

[18] Yanan Wang, Xuebing Han, Dongxu Guo, Languang Lu, Yangquan Chen, and Minggao Ouyang. Physics-informed recurrent neural networks with fractional-order constraints for the state estimation of lithium-ion batteries. *Batteries*, 8(10):148, 2022.

[19] Yanan Wang, Xuebing Han, Languang Lu, Yangquan Chen, and Minggao Ouyang. Sensitivity of fractional-order recurrent neural network with encoded physics-informed battery knowledge. *Fractal and Fractional*, 6(11):640, 2022.

[20] Billy Wu, W Dhammika Widanage, Shichun Yang, and Xinhua Liu. Battery digital twins: perspectives on the fusion of models, data and artificial intelligence for smart battery management systems, energy ai 1 (2020), 100016. *URL https://doi. orq/10.1016/j. ngyai, 2020.*

[21] Ji Wu, Xingtao Liu, Jinhao Meng, and Mingqiang Lin. Cloud-to-edge based state of health estimation method for lithium-ion battery in distributed energy storage system. *Journal of Energy Storage*, 41:102974, 2021.

[22] Yiming Xu, Xiaohua Ge, Ruohan Guo, and Weixiang Shen. Recent advances in model-based fault diagnosis for lithium-ion batteries: A comprehensive review. *Renewable and Sustainable Energy Reviews*, 207:114922, 2025.

[23] Shichun Yang, Zhengjie Zhang, Rui Cao, Mingyue Wang, Hanchao Cheng, Lisheng Zhang, Yinan Jiang, Yonglin Li, Binbin Chen, Heping Ling, et al. Implementation for a cloud battery management system based on the CHAIN framework. *Energy and AI*, 5:100088, 2021.

[24] Qiangxiang Zhai, Hongmin Jiang, Nengbing Long, Qiaoling Kang, Xianhe Meng, Mingjiong Zhou, Lijing Yan, and Tingli Ma. Machine learning for full lifecycle management of lithium-ion batteries. *Renewable and Sustainable Energy Reviews*, 202:114647, 2024.

[25] Guangxu Zhang, Jiangong Zhu, Haifeng Dai, and Xuezhe Wei. Multi-level intelligence empowering lithium-ion batteries. *Journal of Energy Chemistry*, 2024.

[26] Jingyuan Zhao, Xuning Feng, Quanquan Pang, Michael Fowler, Yubo Lian, Minggao Ouyang, and Andrew F Burke. Battery safety: Machine learning-based prognostics. *Progress in Energy and Combustion Science*, 102:101142, 2024.

[27] Yuejiu Zheng, Yifan Lu, Wenkai Gao, Xuebing Han, Xuning Feng, and Minggao Ouyang. Micro-short-circuit cell fault identification method for lithium-ion battery packs based on mutual information. *IEEE Transactions on Industrial Electronics*, 68(5):4373–4381, 2020.

[28] Yusheng Zheng, Yunhong Che, Xiaosong Hu, Xin Sui, Daniel-Ioan Stroe, and Remus Teodorescu. Thermal state monitoring of lithium-ion batteries: Progress, challenges, and opportunities. *Progress in Energy and Combustion Science*, 100:101120, 2024.

CHAPTER 7

Conclusions and Take-Home Messages

7.1 CONCLUDING REMARKS

In this monograph, we have presented fractional-order intelligent modeling techniques for Li-ion battery systems with both theoretical contributions and practical applications using real-world battery data.

In Chapter 2, we presented a comprehensive survey of the state-of-the-art research on fractional-order modeling, including electrochemical and electrical models, parameter identification, and fractional-order estimation methods. The literature review provides an overview about main research directions of fractional-order modeling for LIBs.

In Chapter 3, we explored physics-informed machine learning for battery modeling and SOC estimation with fractional calculus. Battery mechanism shows fractional-order characteristics, which is applied to the design of neural network to make the algorithm informed by battery physic information. Fractional-order gradient and fractional-order constraints are developed for recurrent neural network (RNN), resulting in a fractional-order physics-informed RNN (fPIRNN).

In Chapter 4, we further explored physics-informed recurrent neural network (PIRNN) for battery capacity estimation of realistic running EVs with fast degradation. The interpretability of PIRNN is enhanced by adding ICA inputs related to battery degradation mechanism. Moreover, a battery dataset of ten EVs with a total 5697 charging snippets is constructed to validate the performance of the proposed intelligent modeling algorithm.

In Chapter 5, besides fractional-order modeling, smart sensors are highlighted to enhance signals collected by BMS. Smart perception is introduced, and five types of battery sensors are presented in details. Two example sensors with machine learning are presented. One is a microwave radar array sensor called Walabot for battery voltage classification, the other is a mm-wave radar array sensor called ImageVK74 for battery capacity estimation.

In Chapter 6, an overview of intelligent management system for LIBs is discussed, and perspectives of this intelligent BMS is presented from three main aspects. The

DOI: 10.1201/9781003670902-7

specific functions of fractional-order modeling are highlighted to provide several potential research investigation directions for future work.

7.2 TAKE-HOME MESSAGE OF THE BOOK

In this section, we would like to provide some take-home messages for every chapter, including notes for future work.

1. Chapter 2: Fractional-order elements and fractional calculus are inherently suitable for battery modeling, state estimation, health management, and optimization for battery operation. Fractional-order elements can more precisely capture the electrochemical characteristics of Li-ion batteries.

2. Chapter 3: It is worth investigating how to combine fractional-order characteristics of battery with machine learning, to make the algorithm better informed by battery physics. The proposed fPIRNN is used for SOC estimation of LIBs, while it can also be applied to other state estimation processes (such as capacity, SOH, and RUL) or safety analysis (such as pre-warning and battery pack heterogeneity) in addition to SOC estimations. The optimization of machine learning algorithm by fractional calculus also deserves further research.

3. Chapter 4: For battery capacity estimation, life prediction, or other objectives of LIBs, the corresponding fractional-order constraint plays an important role in constructing an accurate battery model. Hence, fractional-order PDEs needs to be explored to describe different battery characteristics from various aspects, such as electrical aspect, thermal aspect, and physical aspect.

4. Chapter 5: Smart sensors mean more signals and information of battery, which is definitely important for BMS for next generation batteries. However, more adding more sensors may also bring more risk and influence for battery operation, which remains to be investigated and improved. The employment of different types of sensors should be carefully designed. The trade-off of performance and cost needs to be considered. Embedding sensor technology also remains to be improved for future smart perception applications.

5. Chapter 6: Intelligent modeling is a promising framework, and fractional-order modeling should play a vital role in this trend of battery modeling. For cloud and edge computing design, digital twins should be considered. For privacy protection, federated learning should be considered. Smart battery, smart management, and smart control are all based on the "smart design" for battery, and fractional-order modeling should be combined into the design.

Index

For Product Safety Concerns and Information please contact our EU
representative GPSR@taylorandfrancis.com
Taylor & Francis Verlag GmbH, Kaufingerstraße 24, 80331 München, Germany

www.ingramcontent.com/pod-product-compliance
Lightning Source LLC
Chambersburg PA
CBHW080244230326
41458CB00097B/3112

9 7 8 1 0 4 1 1 3 2 6 9 1